VETERANS

VETERANS

STORIES FROM AMERICA'S BEST

PETE MECCA

Nedra,

Hi, Cuz. Thanks for your support & for being you. I wish we could have been together over the years, but life tends to throw a lot of curve balls at us. Wish Mom & Dad was still with us to witness my first book.

You Are Special!

Love
[illegible]

Published by Deeds Publishing in Athens, GA
www.deedspublishing.com

Printed in The United States of America

Cover design by Mark Babcock

ISBN 978-1-947309-22-7

Books are available in quantity for promotional or premium use. For information, email info@deedspublishing.com.

First Edition, 2018

10 9 8 7 6 5 4 3 2 1

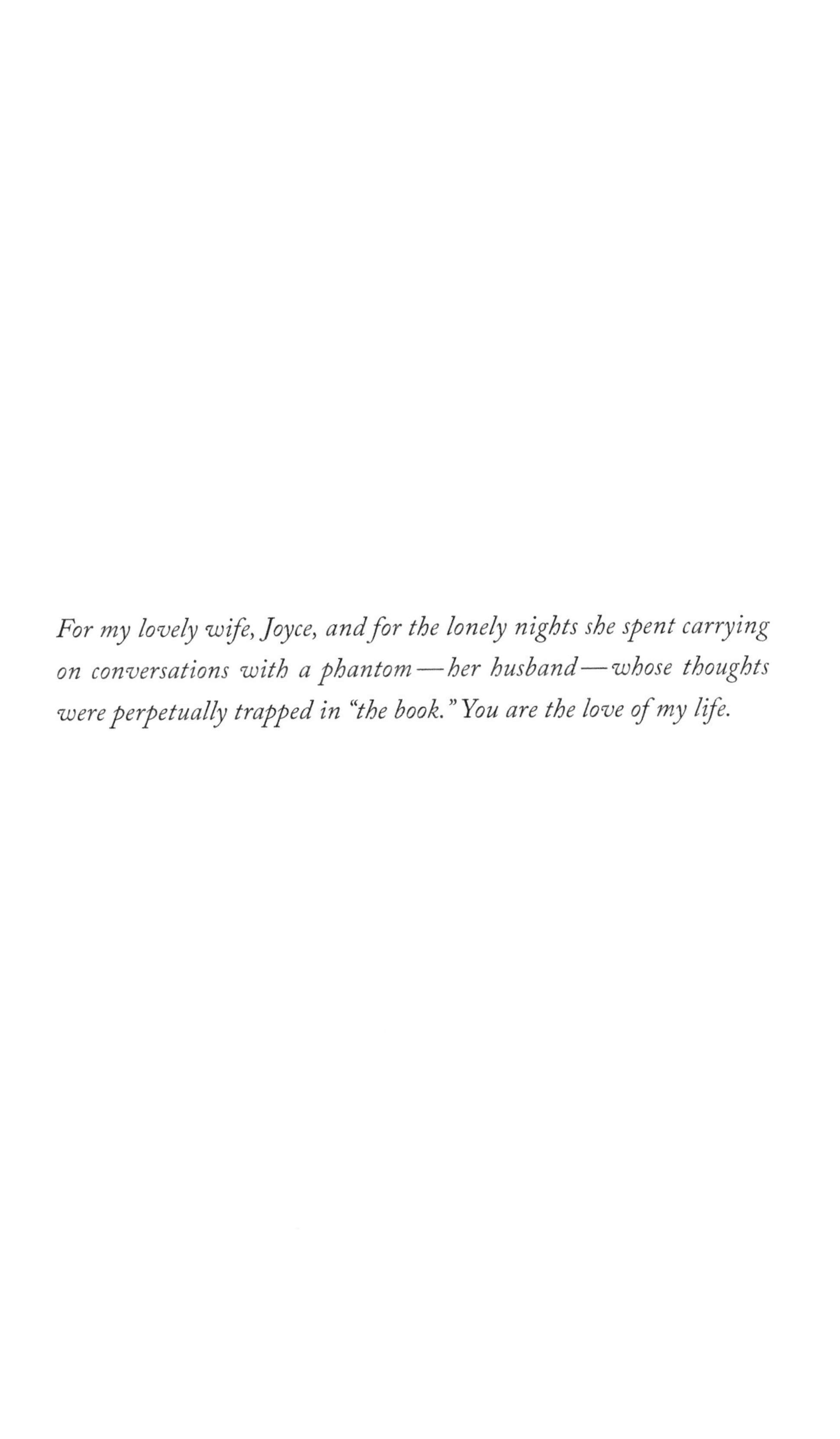

For my lovely wife, Joyce, and for the lonely nights she spent carrying on conversations with a phantom—her husband—whose thoughts were perpetually trapped in "the book." You are the love of my life.

Contents

FOREWORD

> *"We are now in this war. We are all in it, all the way. Every man, woman and child is a partner in the most tremendous undertaking of our American history."*
>
> —President Franklin D. Roosevelt
> in his war broadcast, December 9, 1941

In the not too distant future, while you are watching Fox News, CNN, or another media outlet, you'll be interrupted with the 'Breaking News' that the last veteran of World War II has passed away. Whatever you may think, say, or feel at that moment in history, I hope we can all appreciate the fact that with his or her death the last honest account of WWII has also faded away.

The Greatest Generation is expiring at an astonishing rate, approximately 500 per day as of this writing. With their passing, one adage is irrefutable: America will never again see men and women in the same vein as G.I. Joe and Rosie the Riveter.

The Greatest Generation was a product of The Great Depression. They understood what an authentic food shortage signified because they felt the deficiency in their bellies, plus they witnessed the despondency of family and friends. The frequent promise of a 'middle-class' was more of a pipe dream than the American dream.

Red-necked farmers with calloused hands and Jacks-of-all-trade metropolitan inhabitants worked jobs, any jobs, jobs later generations would consider beneath their dignity. Yet these descendants of hunger and hopelessness retained their faith and remained dutiful to the belief of life, liberty, and the pursuit of happiness.

A summons to arms 'Remember Pearl Harbor' prompted men and women by the millions to don a uniform or belly up to the production lines. Friends and neighbors, school chums, brothers and sisters, wives and husbands, teammates, aunts and uncles signed up and volunteered to support the war effort. It just seemed like the right thing to do.

These citizen soldiers trained as a team. They learned how to kill to prevent from being killed; they drilled together, slept together, ate together, and sailed together across immense, and dangerous oceans. Most of these young men had never been on a ship larger than a rowboat. Many had never seen an ocean. The Great Lakes astonished sailor trainees.

The vessels they boarded were routinely civilian hand-me-downs with a few luxury liners thrown into the mix due to their speed and load capacity. More often than not, boys barfed for the entire journey, many with notions of jumping overboard to end their seasickness. Mississippi Rebels and Damn Yankees from Boston never wanted to board another ship or see another ocean.

In due course, the Liberty ships, prefab production line vessels the G.I.s swore were patched together with duct tape and chewing gum, carried the bulk of war materials.

If not up-chucking from seasickness or camped out in the head with diarrhea from unaccustomed grub, the boys spent their idle time spreading rumors or speculating on what lay ahead. Many of their destinations were unpronounceable, a few sounded exotic, yet these mysterious places were never mentioned in their seventh-grade geography class. Within the coming months that soon converted into combat-hardened years, these strange-sounding names and places would be discussed at every American dinner table, church, and neighborhood bar across the country: Kasserine Pass, El Alamein, Anzio Beach, Hurtgen Forest, Malmedy, Mandalay, Guadalcanal, Peleliu, Rabaul, Eniwietok, and a small sick-smelling sulfur island in the middle of nowhere called Iwo Jima.

They sailed, flew, and marched into combat with southern drawls and Yankee brogues, mid-western clarity and the battle-winning Native American tongues of the Navaho and Wampanoag tribes. Rich and poor whites, underprivileged blacks, Jewish immigrants on the East Coast, and Japanese Nisei from the West Coast, young men of different colors from dissimilar cultures slipped on their combat boots and dog tags, then died by the thousands on distant soil for a dirt airfield, rock-strewn beaches, shark-infested lagoons and tiny atolls, or were forever entombed in the murky depths of Iron Bottom Sound. Young men became old men sprouting white hair in a matter of weeks, perhaps days, many in an hour.

As soldiers battled for mud and sailors fought for saltwater, daring young men in cutting-edge flying machines fought for domination of the airspace over the same mud and same saltwater. Then they redeployed to take one more acre of hedgerow or jungle, a new body of saltwater, or a cloudless blue sky over an even purer blue lagoon. For the men and women who survived, they would never be the same. And back in the 48 states, major economic and societal reform hit Lady Liberty like a sledgehammer. She, too, would never again be the same.

Lady Liberty was in transition, as were her ladies. Close to seven million women worked in defense plants while over a quarter million volunteered for military duty. Lady pilots ferried big bombers and agile fighters across the country, dozens died in accidents. The walls of prejudice revealed their predictable signs of weakness as black Marines proved their mettle on Saipan and the Japanese Nisei battled across Italy and Europe, even as their families endured segregated injustice or were housed as de facto POWs in hastily built internment camps back home.

War tends to boost business. The United States renewed its industrial potency as the Great Depression succumbed to the Great Recovery. Once idle factories, ship yards, and production lines tooled up to unheard of manufacturing levels: 137 aircraft carriers and escort carriers, 806,000 2½ ton trucks, 41,000 guns and howitzers, 12,500,000 rifles and carbines, 41 billion rounds of ammunition, 100,000 tanks and armored vehicles, 310,000 aircraft, and 36 billion yards of cotton textiles. As Japanese Admiral Yamamoto prophesied after assaulting Pearl Harbor, 'I fear all we've done is to awaken a sleeping giant.' Admiral Yamamoto was spot on: Goliath was awake, and Goliath was pissed.

In Washington, DC, political affiliations softened as Hawks and Doves and indifferent Vultures mutated into enraged Eagles with sharpened talons. America went to work and the boys went to war. She was changing, as were her workers, her culture, and eventually the 17 million men in uniform who came home to do their best to live out their lives and the American dream while trying to cope with the nightmares of the war they had won.

*****

> *"This was a police action, a limited war, whatever you care to call it, to stop aggression and to prevent a big war. And that's all it ever was."*
>
> —President Harry S. Truman, speaking on Korea.

President "Give 'em hell, Harry" Truman was known for his bluntness and periodic lack of social graces, but Harry was a down-to-earth type of fella and honest, qualities sorely needed today along the shores of the Potomac. Nevertheless, referring to the Korean War as a "police action" was dogmatic hogwash.

A bona fide "police action" is a swarm of blue uniforms on

a drug bust or responding to disorderly conduct at a frat party or parleying a domestic disturbance with the patience of Job. A 'police action' is NOT 33,665 American warriors killed in action, 3,275 lost to disease, accidents, or crashes, and 103,284 listed as wounded. Unofficially, thousands of American soldiers were wounded, treated by medics, and returned to action: no report made, no Purple Heart received.

To use a catch-all slogan such as 'police action' is an insult to the soldiers and an affront to the families of Korean War veterans. By comparison, my Band of Brothers in Vietnam paid the price of freedom with 58,267 American lives in the 10 years of build-ups, bombing halts, booby-traps, and B.S. from political leaders. That equates to a yearly average of 5,826 flag-draped caskets, and accept my apology for using damnable statistics when discussing American war dead.

To continue, our men fought and froze in Korea for three years, which equates to 11,222 combat deaths each year. Those gruesome statistics are not the statistics of a 'police action' nor do they indicate the suffering. I pray our country never again claims the deaths of thousands of American warriors were casualties of a 'police action', 'a limited war', or the ludicrous politically correct explanation for warfare as an 'overseas contingency operation'—whatever the hell that is.

Google or visit a local library and research the Korean War. You'll notice one habitual phrase: "The Forgotten War." Whoever coined this preposterous phrase obviously never wore a uniform or witnessed the horrors of combat. Maybe the catchphrase delighted a publisher or spawned from the intellect, or lack thereof, of a vain lecturer then caught-on as the exemplar taxonomy of the Korean War. What a pity. The idea that using the word "Forgotten" might offend or dishonor our Korean War veterans wasn't even a postscript.

Korean veterans are in a unique class of American warriors. Their war was wedged between the victorious WWII strategies of 'unconditional surrender' versus a humiliating save face idea of an 'honorable withdrawal' from Vietnam. Their war, from the Pusan Perimeter to the Frozen Chosin, would be the first conflict of many future conflicts to be determined by political squabbling instead of American fire power.

American soldiers won the Korean War. The conflict was over in October of 1950, our soldiers would be home for Christmas, casualties had been light, almost acceptable, yet the world failed to take note of the endless manpower north of the Yalu River: China. Military Intelligence disseminated a plethora of reports that the Chinese intended to intervene on a massive scale. In addition, the Chinese had warned both the United States and the United Nations through neutral embassies that if push came to shove on the Yalu, the Chinese would shove, and shove hard. Sadly and fatally, the reports and warnings were ignored since the war was essentially 'over', the war had been 'won.' If it looks like a duck, walks like a duck, and quacks like a duck; it's a quarter million Chinese soldiers waddling across the frozen Yalu River.

An anxious President Truman and the hard-headed General Douglas MacArthur traded conflicting remarks concerning strategy and objectives as American soldiers and Marines fought for their lives in minus-zero temperatures during a disorganized-organized retreat. Or as the Marines claimed, 'We're just attacking in a different direction.' Soldiers died from untreated wounds, heroic but exhausted Marines froze to death in defensive positions, they ran out of ammunition, food, and hope. They weren't going home for Christmas.

Ultimately reorganized and reequipped, America and her allies regained lost ground, lost the field again, then battled back for desolate hill tops with names like Bloody Ridge, White Horse, Old Baldy, Jane Russell, Heartbreak Ridge, and Hill 122, a piece of worthless real estate the Marines nicknamed Bunker Hill. As seesaw battles continued, young men on both sides breathed their last breath holding or trying to retake precarious pieces of real estate along the infamous 38th parallel, while alleged peace talks dragged on at Panmunjom.

Military Intelligence is frequently teased as an oxymoron, but the Korean War 'peace talks' at Panmunjom were assuredly the ultimate oxymoron. Negotiators heatedly debated the size of the conference table and the dimensions of national flags. The Communists accused Americans of engaging in germ warfare, without proof, an accusation so absurd the Soviet delegate at the United Nations chose to remain silent on the issue. So the fighting continued. Men fought for a few extra yards of bloody earth, useless real estate traded for human lives so negotiators could gain the political 'edge' at a conference table.

This new type of warfare did not sit well with the American

people. We fought and won a global war so what the heck is taking place in Korea, and why are we still there? Difficult questions were asked but solutions were in short supply. Interest in the war faded away quicker than the old soldier in General MacArthur's famous 'fade away' speech before Congress. The fighting continued, and men died as other men with bantam rooster personalities bickered about a word or a comma or a paragraph, or the bloody communist POW riot at Koje-do prison.

The madness eventually ended with the cease-fire agreement signed on July 27, 1953, which was nothing more than an unadorned armistice between two militaries agreeing to stop the killing while both sides sought a political solution. Not much of a victory dance for the American boys we lost in combat. Officially, we are still at war with North Korea, thus all the saber-rattling and swagger from their twerp-like leadership.

So the boys came home. No victory parades, no stately victory speeches, nobody cared. Over sixty years later, the grandsons of Korean War veterans are still there, guarding the 38th parallel and praying to God that unreliable diplomacy will keep them safe from twerps with nuclear weapons.

* * * * *

Approximately 10 months after the cease-fire was signed in Korea, French Colonel Christian de Castries announced via radio that his stronghold at Dien Bien Phu had fallen to the Viet Minh. French colonialism had died a long-drawn-out death in Indochina, or as the locals called it, Vietnam.

The United States increasingly filled the void by sending advisors and Green Berets into South Vietnam, Laos, and Cambodia to hopefully stem the tide of world communism. The ensuing brouhaha developed into a 10,000-day war. The first recognized American casualty, WWII navy veteran Richard B. Fitzgibbon, who later joined the Air Force, died of his wounds on June 8, 1956 at the age of 36. On September 7, 1965, his then 21-year old son, Marine Lance Corporal Richard B. Fitzgibbon III, was KIA in Quang Tin Province. Both names are etched into the black granite on the Vietnam War Memorial in Washington, D.C.

"War would end if the dead could return."

—Stanley Baldwin

* * * * *

The CBI Theatre of Operations (China, Burma, and India) kept my father overseas for almost three years during WWII. That was his war. My war kept me in Southeast Asia for 30 months. Like father, like son, but in two different conflicts. My father knew for what he fought: Freedom, and the 'unconditional surrender' of our enemies. His son only thought he grasped the logic behind the Vietnam War. My generation of baby boomers grew up on the stories of WWII, read the books, and cheered John Wayne as he and fellow Marines took Mount Suribachi in *Sands of Iwo Jima* (Three of the surviving flag-raisers appeared in the movie). To baby boomers, Korea was a vague concept of something called 'limited war' which was misunderstood by the majority of American families. Except for kids like Steve Burke, my high school buddy, whose father was killed fighting for an eventual stalemate on the peninsula known as 'The Land of the Morning Calm.'

Exactly what happened in this mysterious and secretive third-world country called Vietnam? What caused half of my generation to rally around the flag while the other half burned it? In the end, a truism prevailed that even the most powerful country on earth cannot arrogantly trample across quicksand.

Half-ass wars result in half-assed results. The price of such folly can be reviewed on the 58,307 names etched into a 494-foot black granite wall, a wall void of elaborate statuary and without an inscription to identify which war it represents. America lost its innocence in Vietnam, much like a young virgin regretting a bad decision. The psychological effects of that decision still walk our streets today.

My personal journey home from Vietnam came to fruition on Veterans Day weekend in 2011. I was fortunate to be on one of the four airliners provided by the History Channel which flew Vietnam veterans into our nation's capital to enjoy three days of whirlwind activities, including a special ceremony at The Wall. Joe Galloway, the famous writer-journalist who remained on the ground throughout the battle of the Ia Drang Valley, was our guest speaker. Joe did not pull any punches, he spoke the truth, and we embraced his veracity. After nearly 50 years of nightmares and memories, I was finally able to approach the chunk of black granite and caress the names of guys I knew a long time ago.

One name etched into the black granite was Douglas Reyes. Doug and I graduated in 1965 from Bartlett High School, just northeast of Memphis, Tennessee. Doug, rough and tough, joined the US Marine Corps. As a student pilot behind the controls of Cessna 150's and 172's, my flying passion channeled me to the US Air Force. Doug and I both ended up in a country, unlike Indonesia (Emerald of the Equator), Japan (Land of the Rising Sun), Thailand (Land of Smiles), or Sri Lanka (India's Teardrop), without an internationally recognized nickname. This land, this war-torn piece of real estate under 2,000 years of foreign domination, wasn't even officially recognized by the name "Vietnam" until 1945.

The G.I.s had slang words for Vietnam, many unprintable, but Doug and I ended up in this no-nicknamed land of land mines and leeches to fight in someone else's civil war. It just seemed like the right thing to do. As I stepped on and off airplanes, Doug stepped on a land mine. He was 18 years old. Over 40,000 names on The Wall were 22 years old or younger. Pfc Dan Bullock was the youngest, he was 15.

Was Vietnam worth the price? If one considers the fiascos of a war run from the basement in the White House, absolutely not. But Pfc Dan Bullock, Marine Doug Reyes, and Peter Jay Mecca Sr.'s flyboy progeny had our war, it was our time to serve, and we did not shirk from our duty.

As a Vietnam veteran and free-lance writer, the literary road less traveled would produce a repetition of the heroics of our warriors and their hard-fought battles: the 1st Air Cav in the Ia Drang Valley; the infamous Tet Offensive, Marines clinging to and eventually winning the siege at Khe Sanh; F-4 Phantoms, F-105 Thunderchiefs, and F-8 Crusaders dodging lethal surface-to-air missiles and murderous anti-aircraft fire going 'downtown' to Hanoi, or I could rehash The Battle for Hue. Any critique on the combat in Vietnam, Laos, and Cambodia should include the Pentagon Papers, egocentric politicians, war protestors, and an actress turned political activist named Jane Fonda, but I'll leave those rudiments of the war to historians and heretics.

My job is to tell the truth. Two thirds of the American men and women in Vietnam were volunteers. We joined, we served, and mercifully most of us survived. We withstood and endured things John Wayne never filmed nor could he film, having never worn the uniform. We, the military lifers and the fresh meat baby boomers, believed in baseball, hot dogs, apple pie, and Chevrolet, with several Fords thrown into the garage to keep the economy in sync. Yet we spent our formative years with leeches, dung-filled rice paddies, impenetrable jungles, cobras, punji stakes, jungle rot, and slant-eyed adversaries who all looked the same. We even developed a fondness for French Cognac.

We compared war to football: gear-up, take the field, kick butt, and then go home. Not in Vietnam. We kept our boys on the field too long, drained their spirit, and brutalized their strength of character. American players were ordered to block, pass, and tackle with one hand tied behind their backs while the opponent played by their own rules, never saw a penalty flag, and covertly recruited cheerleaders, concession workers, and bleacher bums to kill us on the sly. Foul play by the other side, no matter how abhorrent, could not be challenged and replays were prohibited. The game was rigged.

The Greatest Generation vanquished the Axis Powers then sailed home victorious. The catastrophe in Korea was, well, forgettable, a draw, not fought by the book and 'limited' in scope, but America learned a valuable lesson at Panmunjom and on dozens of blood-soaked barren hills. The old adage, "*In war there is no substitute for victory,*" attributed to General Dwight Eisenhower would be remembered and adhered to in future wars. Or so America thought.

Lo and behold, a US Navy destroyer was attacked on the high seas by North Vietnamese torpedo boats. A second destroyer reported an unconfirmed attack, and in a flurry of passion the US Congress passed the Gulf of Tonkin Resolution which gave President Lyndon Johnson the no-holds-barred authorization to wage an undeclared war in Southeast Asia.

We fought 10,000 miles from the 'real world' of round-eyed women with blonde hair (bottle-crafted in some cases) in exchange for endless jet black hair. We actually *were* in another world, another culture, inhabited by another race, embroiled in their civil war with no end in sight except for our determined enemy. American soldiers were ensnared in a no-win situation; Washington knew it, but lacked the guts to call a duck a duck. If a nation does not learn from its history, then bad history will repeat itself. Iraq and Afghanistan come to mind.

American boys fought under awkward rules of engagement. But a soldier is a soldier, and soldiers are in every culture, every clan, every tribe, every country; and once a warrior, always a warrior. When warriors return from combat they cannot simply turn off their experiences like a light switch. Native American Indians accept the truism that returning warriors need time to heal their 'spirit' after combat, plus Native American warriors are shown respect and honor by their tribes. This is not the case in Biloxi or Baltimore or Boston. We, as the greatest country on earth, did a damned lousy job transitioning our Vietnam veterans back into civilian life.

Post Traumatic Stress Disorder (PTSD) is today's fashionable lingo for previous catchphrases like shell-shock, battle-fatigue, or trench-disease. They all mean the same thing: a warrior is psychologically scarred by war. PTSD grew in acceptance in the years following Vietnam, but the trauma has been customized and molded into categories, subtitles, and visual aids analogous to sentence diagrams in English 101 to where hangnails qualify as PTSD. Warriors in desperate need of medical attention are judged ineligible for a myriad of rationale, yet during the War in Vietnam one airman stationed on Guam was granted 100% PTSD disability claiming loading bombs on B-52s 'traumatized' him when thinking of the people on the receiving end of the payload. He never set one polished combat boot in Southeast Asia. I've interviewed dozens of veterans beset by PTSD. Their credibility cannot be denied. But the others, well, a fake is a fake, a con is a con, and a duck is a duck.

One false concept concerning Vietnam is that we 'lost' the war. If 'we' implies the men in uniform, then the statement is totally deceitful. If 'we' means the decision architects in Washington, DC, then truth wins the day. The American soldier in Vietnam never lost a major battle.

The commander of the North Vietnamese Army, General Vo Nguyen Giap, has repeatedly been quoted as saying the US had the war won twice: After Tet '68 if we had pushed into North Vietnam, and in the course of the B-52 assaults during the 'Christmas bombing' in 1972. Conceivably, both would have worked and stopped the war, at least for a few years, but Giap did not make those two statements. General Giap also DID NOT attribute the 'anti-war' movement to be an important or significant part of the NVA's victory in 1975. A retired NVA colonel made the claim concerning the 'anti-war' movement.

General Giap was an exceptional student. In his early days, Giap worked as a messenger for a power company, taught history, joined the Communist Party, penned revolutionary essays as a journalist (Ho Chi Minh often criticized Giap for his verbose writing style), fought for Vietnamese unity and climbed the ranks of the military elite. His strong suit was the concept of *Protracted War,* basically, dragging a war out until an enemy capitulates. Although the theory is largely attributed to Mao Zedong, Mao heavily paraphrased 'The Art of War' by Sun Tzu. Giap, following the example of the first Premier of the People's Republic of China, Zhou Enlai, skillfully expanded one sector of *Protracted War* identified as *Sparrow Warfare*, meaning a fast gathering and fast dispersal of inferior forces until strength and manpower develops int[illegible] force capable of engaging an enemy in conventional co[illegible]

Do these tactics work? One would ha[illegible]ge Washington or General Charles [illegible] taking place on the stre[illegible] the puppeteers pu[illegible] professors and restric[illegible] on the unruly actions of activist-[illegible] in Washington, D.C. itself. To underestimate '*p*[illegible]*a or sparrow warfare*' is to flirt with disaster.

As the mastermind of the French defeat at Dien Bien Phu, Giap was perfectly willing to accept limitless casualties to defeat his enemy, although he was firmly against the Tet Offensive. But in the end, he and the leadership in North Vietnam again rolled the dice in South Vietnam with a country-wide attack remembered as the Tet Offensive of 1968. It totally failed. Conventional warfare with overwhelming firepower was Uncle Sam's gridiron. Giap missed his extra point, but he did score significant points in the bleachers: Giap's gamble broke the tenacity of the American people and their representatives in Washington, D.C. He had won the psychological victory.

Then came the year of the BUFF's; the United States Air Force's huge B-52 bombers. In 1972, they winged over North Vietnam to drop enormous payloads on the enemy's homeland. Power equates to punishment. Fearful of total destruction, the North Vietnamese decided the best avenue to victory lay in the stalled 'peace talks' in Paris. Shades of Panmunjom.

Monday morning quarterbacks can quarrel and debate and promote their analytical intellectual judgments concerning Vietnam, Iraq, or Afghanistan until their faces turn blue. These people lack combat experience or military training to even make a difference plus make a lot of money verbally fighting old battles. The facts are undeniable: Never send American warriors into no-win situations; never hold back the fire-power required for a prompt conclusion; and you negotiate from strength, not weakness. And fight to win.

American warriors are football players; give them the ball, let them run with it, they'll kick butt, then get our boys the hell home.

* * * * *

Operation Urgent Fury, the American Invasion of Granada in 1982, a tidy little war on a tiny little island. One fight in which Goliath slew David. Big and bumbling, it quickly became clear that Goliath needed to upgrade his tactics and strategy before weapons as crude as a slingshot hit him squarely in the forehead. The outcome was never in question, but the 'win' should and could have been a fairly simple endeavor instead of a Keystone Cops operation that cost American boys their lives.

Coordination between different branches of service was poor. Communication was amateurish. American soldiers had to use tourist maps with military grid references drawn by hand. 'Friendly fire' episodes occurred due to lack of synchronization. One soldier called long distance to relay the need for close air support. The topography of the island was unfamiliar. Albeit, Goliath upgraded his entire contingency planning.

By the time Saddam Hussein decided Kuwait was ripe for the picking, the American military machine was, by a wide margin, the most daunting, well-equipped, and well-coordinated football team taking the field. Their able coach, a Vietnam combat officer and Deputy Commander of the Granada Invasion, General Norman Schwarzkopf, also had an impressive new playbook.

* * * * *

"Get there first with the most men."

—Attributed to Gen. Nathan Bedford Forrest.

'Get there first with the most men' is frequently misquoted as, '*Get there firstist with the mostist.*' Saddam Hussein charged head first into Kuwait firstist, but Uncle Sam and a coalition of allies followed suit with the mostist. In this conflict, Goliath was called upon to save David from the Wicked Witch of the North.

President Bush (the older guy) diplomatically super-glued a maverick coalition of mismatched political, religious, and social dissimilarities into a cohesive fighting force with one goal in mind: kick the bastard out of Kuwait. Much like Vietnam, we won and we lost.

Forcefully kicking the Iraqi military out of Kuwait (Desert Storm) is conceded in history books as an overwhelming victory for the allied coalition. Our smart munitions and smart warriors, led by a smart general named Schwarzkopf, were allowed by a WWII combat veteran President George H. W. Bush to do what they had been trained to do, then Johnny Came Marching Home.

A proud country watched and thousands attended an inspiring victory parade in Washington, DC. Respect for the military had returned, and Vietnam was forgiven as a bad memory. Not one 'protester' made the 6 o'clock news burning an American flag. Kudos poured in for Bush. Schwarzkopf's no-nonsense tactics and strong personality earned the general a slot as an American military icon and his book, *It Doesn't Take a Hero,* became an overnight best seller. Yeah, we won a war, decisively so, but we lost the prize: Saddam Hussein was still in power, and the head of the snake still had fangs.

* * * *

The Panama Canal, Kosovo, an incursion here, a firefight there, maintaining the peace and staying strong, so strong we could take a well-deserved national nap. February 26, 1993: The 'sleeping giant' was rudely awakened by a massive truck bomb detonated beneath the North Tower of the World Trade Center. Six dead and over a thousand wounded. An irrelevant piss ant terrorist group called Al-Qaeda claimed responsibility and had the gall to 'declare war' on the United States. Yeah, sure, as if a piss ant could do us any harm.

Time is on the side of terrorism. On August 7, 1998, the 8th year anniversary of the arrival of American forces in Saudi Arabia, the United States got a glance of things to come. The American embassies in Dar es Salaam, Tanzania and Nairobi, Kenya had been reduced to rubble. The butcher's bill: 258 dead, over 5,000 wounded. Perhaps that little piss ant was serious. In response, President Clinton ordered a cruise missile strike on Osama bin Laden and his Al-Qaeda training camp in an isolated area of Afghanistan. The missile strike impressed a couple of camels.

As 1999 retired a century of war and bloodshed, the New Year 2000 exposed a plot to hijack ten commercial airliners at one time. The plot thwarted, President Clinton and other politicians congratulated each other and gave the Intelligence community a pat on the back. America could now rest easy knowing their country was on top of terrorism; we had stepped on the piss ants. While we applauded a limited victory, thousands of other piss ants poured out of their secretive mounds. The worst was yet to come.

During early 2001, strange reports started arriving at various

American Intelligence Communities. Apparently middle-eastern males were entering flight simulators to learn how to pilot commercial airliners...straight, no interest at all in take-off or landing procedures. Trained to disseminate information and produce a proper conclusion, the folks in Intelligence fumbled the ball on the one yard line. Nothing was done and few in power even showed interest.

September 11, 2001: The piss ants roar. The towers tumble down. Death visits the East Wing of the Pentagon while an airliner is reduced to a million smoking pieces in a vacant field in Shanksville, PA. Four commercial aircraft had been hijacked and used by middle-eastern men as guided missiles to attack the United States of America. I suppose the ill-trained pilots didn't have much of a choice since they didn't know how to land.

Let's rehash what took place, then maybe we can begin to understand exactly what happened. A group of terrorists declares war on the United States of America. They are well-funded, well-armed, disciplined, full of ideological zeal, and loathe capitalism and Christianity. They receive megabucks from oil-producing sympathizers, feel the need to target men, women, and children, do not and will not follow any rules of engagement (sound familiar?), and are trained to do only one thing: kill us. Here's a thought: 'declare war' on the piss ants, just to make things legal, and destroy their mounds before they destroy ours.

Reenter politics. The towers are gone, the nerve center of our military severely damaged, and a field in Pennsylvania littered with airplane parts. We invade Iraq then invite Saddam to a necktie party. His neck is stretched, and that's a good thing, but we should have hung the butcher after the Gulf War. We get out of Iraq, we go back into Iraq; and Afghanistan, a land known as 'The Graveyard of Empires', continues to gut our resources. It took us ten years to put a bullet in Osama bin Laden. Well, several bullets. So now it's time to come home. Nada. We're still there, with no end in sight. In the scheme of things, the foolish idea of a 'free' or 'capitalist' Afghanistan is not worth the life of one American soldier who, once again, is fighting under impossible *rules of engagement* that guarantee the loss of young lives. An '*overseas contingency operation*'—not.

In spite of all this madness, the American warrior continues to saddle up to do his duty while history repeats itself as the men and women of the United States military fight with one hand tied behind their backs. These warriors hang tough. They hunker down, kick butt when allowed, and at least for now are honored and respected for serving their country on another mission impossible. They will come home, and I pray to God Almighty that the political animals will not turn on them for a vote nor cheat them out of the benefits they have earned.

Yes, our warriors will come home, and like my Band of Brothers from Vietnam their transition to civilian life will be difficult. I've interviewed young people who served in Iraq and Afghanistan, but most of them will not talk, not yet anyway, they can't. There's a heartsick feeling down deep in their souls that will linger for years to come, perhaps a lifetime. Like millions of warriors of the Greatest Generation, their stories will be buried with them. It will be up to us to understand the depth of their sacrifice and to be sure that the next war is not just another exercise in political folly.

This book tells the tale of American warriors, the human beings who fought and suffered and died to preserve what other warriors attained for their generations: the simple yet enduring concept called FREEDOM. A foreign military will never invade, occupy, and strip us of our freedoms. Albeit, Abraham Lincoln recognized how America could fall from God's grace: *"America will never be destroyed from the outside. If we falter and lose our freedoms, it will be because we destroyed ourselves."*

Then you will know the truth, and the truth will set you free.
—John 8:32

Not meaning to be so presumptuous as to rewrite the Good Book, but perhaps more fitting for the uninformed of today:

Then you will know the truth, and the truth will KEEP you free.

Enjoy the stories.

THE GREAT WAR: WORLD WAR I

In a wood they call the Rouge Bouquet, There is a new-made grave today, Built by never a spade nor pick, Yet covered with earth ten meters thick, There lie many fighting men, Dead in their youthful prime, Never to laugh nor love again, Nor taste the Summertime.

—Joyce Kilmer, "Rouge Bouquet"
March 7, 1918

THE GREAT TRAGEDY

On July 28, 1914, Austrian Archduke Franz Ferdinand and his beloved wife, Sophie, were in Bosnia en route via motorcar to the residence of the provincial governor in the provincial capital of Sarajevo. The couple traveled under a threatening cloud of danger. Despite repeated warnings and indications of possible violence by Serbian Nationalist members of *Narodna Odbrana* (National Defense) and a clandestine offshoot called '*The Black Hand*', the risk to the royal couple was either downplayed or simply ignored.

An assassin had tossed a grenade in the direction of the couple earlier that morning but exploded behind their vehicle, injuring the occupants in the following motorcar. Undaunted, the royal couple motored on, completed their visit with the provincial governor, then demanded to be taken to the hospital to visit with the injured men from the grenade attack. As fate would have it, the drivers were not informed of the itinerary change and took the wrong route. Once the mistake was realized, the royal couple's driver backed up onto a side street where the motorcade came to a halt.

One of the six members of an assassination squad was seated at a café across the street. Nineteen-year-old Gavrilo Princip jumped at the opportunity. He crossed the street with revolver in hand and opened fire. The Archduke died within minutes; Sophie expired en route to the hospital.

The cause of The Great War, later known as World War One, is generally blamed on the assassination of the Archduke and his wife. True, their murders ignited anger and sabre rattling from numerous ethnic groups, kingdoms, countries, and suspicious religious and tribal alliances. But the little discussed catalyst for The Great War can be found in the turn of the century military concept called 'Mobilization.'

In 1914, the leaders of the era were not blessed with a 'hot line' or any other quick avenues of communication. So-called rapid transportation was a smoke-belching train navigating perilous mountain passes or a rattletrap car bouncing along primitive deeply-rutted roads. Airplanes were considered a newfangled invention without a significant future as a dependable means of transportation. Carrier pigeons were still winging messages to and fro but by the time the pigeons flew their coop and landed on another coop with any 'consultation' overture wrapped around their skinny legs, the fog of war had changed dramatically.

Dewey Garton, middle, in WWI

To protect borders, citizenry, and their government, the military machines of the era began what they had practiced for years: *Mobilization.* An unstoppable buildup of men and materiel was moving into positions, primed and ready.

These were armies of unreliable trucks and airplanes, sluggish trains, antediluvian generalship, and millions of horses. Ornate and flamboyantly dressed cavalrymen armed with lances would soon spur their steeds recklessly into nests of machine guns and die by the tens of thousands, including their steeds. Food, ammo, artillery, uniforms, hay, saddles, shoes, and horseshoes, medical equipment, gun belts, canteens, millions of officers and men, all had to come together as fast as possible under near-impossible criteria. They had to *mobilize.*

When one panicky government heard through rumor mills or political grapevines that another country was *mobilizing,* thus the urgent need to 'mobilize' their own forces or be caught with their military pants down. Like an army of dominos, the military machines fell in line until the officers and soldiers, even their horses, were at long last ready to fight. Problem was, on what battlefield?

The German military was governed by the Schlieffen Plan, the brainchild of Count Alfred von Schlieffen: Use a huge wheeling maneuver through neutral Belgium into the French frontier. The men and materiel needed for the plan to succeed would be enormous and require total loyalty to an over-whelming 'right flank' stratagem. Schlieffen's last words on his death bed confirmed his obsession: "…keep the right flank strong."

Germany hit first; The Great War was on. A modified Schlieffen Plan also kept the 'left flank' strong, thus robbing the Germans of their much desired swift victory. As Alfred von Schlieffen turned over in his grave, millions of soldiers dug thousands of miles of trenches and began dying by the thousands, then into the millions.

The Great War was a massive killing field that could have been prevented, but the lack of communication and common sense turned the conflict into The Great Tragedy.

GENTLY INTO THE GOODNIGHT

Passchendaele is one of many low elevations approximately five miles east of the Belgian town of Ypres. On that gentle rise is Tyne Cot, the final resting place for thousands of young English soldiers. Twelve thousand boys are buried at Tyne Cot, victims of the 3rd Battle of Ypres. The cemetery also commemorates the 35,000 American doughboys that were never found after the battle. Too, a substantial monument at Theipval, France records the names of 73,000+ missing soldiers from the battle of the Somme.

Casualties of that magnitude would be totally unacceptable in today's world of wars, yet the horrible killing was normal and, unfortunately, expected during The Great War. The Great War was more commonly called World War One after Hitler invaded Poland in 1939 to instigate World War Two.

Over 65 million military participants fought in World War One. Of those 65 million participants, nine million combatants would perish, along with seven million civilians. The last living veteran of World War One, a British subject named Florence Green, died on Feb 4, 2011 at the age of 110. An era had slipped gently into the good night.

THE DIARY

"In case I am killed, please send this diary to Mrs. O. J. Garton at 4416 Grand Avenue West in Eau Claire, Wisconsin, U.S.A."

Thus began the diary of World War I veteran, Private Dewey J. Garton, Battery "F" 61st ARB C.A.C. (Field Artillery). His entries start on June 14, 1918 and culminate upon his safe separation from service on Feb 25, 1919. The following excerpts articulate the carefully written verses of a world tragedy as penned by Private Garton, who realized full-well those words may be read by a grieving mother.

June 14, 1918—less than three weeks into his 19th birthday, Garton's unit departs Fort Siever, GA for Camp Eustis, VA.

"We were a happy bunch and anxious to get started."

July 11, 1918—Leaving Camp Eustis, the soldiers march to a small stream to board motorboats for the crossing. Once on dry land, the men march through Newport News, VA to Camp Stuart.

"We went into quarantine."

July 17, 1918—Following a six-day quarantine, the soldiers marched back to Newport News to board a ship for France.

"The Red Cross cheered us by giving us hot coffee and cake before

boarding. I will never forget my last step on American soil. I hate to leave it, yet I want to."

July 18, 1918—10:00am: The convoy sails with an escort of two torpedo boats and one cruiser.

"....finally into the Great Atlantic. Oh my! What a feeling or sensation. The big ship sailed along so smooth and fast."

July 20, 1918—The balance of the convoy gathers near Hobopin, NH and sails for France.

"....the sea was somewhat rough. The big ship would rock and roll. It was a lot of fun, if a fellow could stand it. We sailed away together, almost 20,000 strong."

July 30, 1918—The convoy is bushwhacked by a German submarine.

"I was lying in my bunk about 7 o'clock reading when all at once the alarm bell began to ring. Some were excited and scared. All at once "Bam!" and the ship shook madly. I thought a torpedo had hit us but it proved to be one of our six-inch deck guns shooting at a sub close by."

Pvt. Garton sprints topside to observe a sea battle as the escorting destroyers engage German submarines. The destroyers feasibly sunk one submarine before the other subs skedaddled.

"We were as calm as could be and laughing and talking. We sailed into port at 8 o'clock that night. The people on shore waved handkerchiefs and hands at us. The French people seemed comical. Their talk was more so. I think I will like France just fine."

August 7, 1918—Garton boards a train near St. Nagier.

"It was very crowded and uncomfortable. That night I slept on a plank a foot wide and certainly did not like it."

August 8, 1918—*"I never suffered so in all my life as I did last night trying to sleep! I prayed for morning to come."*

August 15, 1918—*"....took a long hike with gas masks on. Went several miles with no stopping. We were nearly exhausted."*

Throughout the diary, Garton makes references to his gas mask. Chemical warfare in WWI was deadly and painful. The Germans started all the foolhardiness on April 22, 1915 by firing in excess of 150 tons of Chlorine gas on French troops in Ypres, Belgium. The Allies, including the Americans, soon followed suit. Agents included tear gas, mustard gas, phosgene, chlorine, and several more that caused swelling of the eyes, nose, armpits, and any soft skin area. Blindness, massive blisters, severe headaches, and pneumonia caused by blisters on soldiers' lungs hospitalized tens of thousands. Thousands more died a horrible death.

Training, letter writing, marches, signal school, parades for V.I.P.s, bunk and field inspections, even sightseeing became daily routines for Garton.

Sept 26, 1918—*"Went to old grave site now being dug up. ...lots of skeletons and bones. Found hair combs and hair, too. It certainly was gruesome."*

Oct 2, 1918—*"Went through tear gas chamber. Took our masks off in room. Gas attacked my eyes...made tears run and pained extremely."*

Rain, cold, mud, disease, firing artillery, and, yes, American football and baseball in the rear areas between the killing times, were common.

Oct 21, 1918—"...gas instruction by Lt. Gee in the gun pits. I certainly don't want to be gassed."

Oct 23, 1918—*"Tractors returned the 61st guns to Libourne this morning. The names of our four big guns of Battery F are: Weeping Willie, 'Black Jack', U.B. Damn, and Big Ben. We are also using the French 75's."*

Oct 28, 1918—*"...the 61st Artillery is moving to Troyes. It is a city lately inhabited by Germans. It has now been captured by the Americans."*

Oct 3, 1918—*"Good war news. The Kaiser has abdicated. Austria is begging for peace and Turkey is solely in the power of the allies. Wrote a letter to Calla. I sure wish I was with her now."*

Nov 11, 1918—The Great War is over. *"Today will be a memorial day in the history of the world. We joined the main battery with gas masks, helmets, rifles, and side arms, just in case. The Kaiser has fled to Holland in a powerful motor car. The King and Queen of Bavaria have fled from the throne. The Crown Prince of Germany signed away all rights to the throne. The German fleet refused to act. The sailors hoisted Red Days of Revolution. The officers were shot and killed. In many cities, the revolution has broken out."*

As The Great War ended, guns fell silent and soldiers stopped killing, but misguided political seeds were sown to guarantee a future conflict known as WWII. Communists eventually governed Russia; the Empire of Japan was secure as a Pacific power, Britain's colonial rule had stared to fade, and the paranoid French initiated a building project, squandering millions of francs on the useless Maginot Line. And a scrawny German corporal named Adolf Hitler vowed to revenge his country's defeat and humiliation.

January 30, 1919—Coming home.

"We steamed out of Marseilles at 5:00pm. I ate supper and went to bed. I slept real good."

February 25, 1919—The last entry. *"The boys left Camp Upton about 8:00am. It was a sad separation."*

As I browsed Garton's war diary it became apparent to me that this young American doughboy had dampened the words of his combat experiences and the unbelievable horrors he witnessed on the Western Front, and for good reason. Had he not returned, the diary would have been sent to his mother. Wisely, had he died, this good son did not wish his bereaved mother further sorrow.

Garton served in World War Two, teaching gunnery in Texas to the next generation of soldiers. When asked why, at 44 years of age, he was again in uniform, Garton replied, "I was drafted."

My deepest appreciation to Mrs. Kimberly Johnson for trusting me with her very special family heirloom: her grandfather's WWI diary.

WORLD WAR II

"There is one great thing you men will be able to say after this war is all over and you are at home once again. And you may thank God for it. You may be thankful that twenty years from now when you are sitting by the fireplace with your grandson on your knee and he asks you what you did in the Great World War II, you won't have to cough, shift him to the other knee and say, 'Well, your Granddaddy shoveled shit in Louisiana.' No, sir! You can look him straight in the eye and say, 'Son, your Granddaddy rode with the great Third Army, and a son-of-a-bitch named George Patton.'"

—General George Patton
in a speech to troops of the
Sixth Armored Division, May 31, 1944

DOROTHY THE WELDER VS. ROSIE THE RIVETER

I first met Dorothy Turner during an interview with another WWII female veteran at the Morningside Assisted Living Facility in Conyers, Georgia. Seemingly shy at first, once Dorothy started talking you shut up, sat back, and listened. The lady spoke with conviction, she spoke the truth, and she expected to be heard. Her tone was not belligerent or even threatening, yet Dorothy spoke in a voice best categorized as the soft bark of a United States Marine coiled to strike if provoked beyond approval. Later in the week, I received a call from her son, Dan Turner.

U.S. Marine, Dorothy Turner

Dan asked if I'd be interested in interviewing his mother, a WWII Marine veteran that you did not want to aggravate, disrespect, or sweet-talk unless you aspired to have the holy crap beat out of you. His request sounded enticing, to say the least.

I interviewed Dorothy in her modest room at Morningside with Dan sitting beside his mother in case there was a need for recall assistance or feasibly statement clarification. In all honesty, it was Dorothy that corrected both of us, more than once.

Afflicted with the aches and pains attributed to the Golden Years, Dorothy's razor-sharp mind and humorous critique remained intact. She was forthright, not of rudeness, but of a woman trained as a Marine with no time for poppycock. Her values were that of a Marine, and she expected respect be given for the respect she tendered.

A Marine photo exhibited in her room reflected the image of an attractive, energetic, patriotic woman with a smile to enchant any man. Sadly, the Golden Years had taken their toll, raiding her beauty and her youth, but they had failed to diminish her lovely smile, her equanimity, nor her stature as a United States Marine.

Her husband served in WWII as a Navy Seabee at locations few Americans had ever heard of until the victories and casualties on distant shores became front page headlines, places like Guadalcanal and the Aleutian Islands. Dorothy and her husband endured the hardships of WWII yet lost their son, Michael, in the Vietnam War, fighting the crucial battle to retake the ancient city of Hue during the 1968 Tet Offensive. Michael had been 'in-country' for one week.

"A woman is the only thing I am afraid of that I know will not hurt me."

—Abraham Lincoln

Obviously, Mr. Lincoln never met Dorothy.

Dan Turner is always apprehensive when he receives a call from the staff at Morningside Assisted Living Facility. His mother, Dorothy, resides at the facility and has not been in the best of health as of late. Dan said, "I received a call one evening from Morningside and was told I'd better get out there as quickly as possible. Of course being concerned, I asked them what was wrong. They told me, 'The Marines have landed at Morningside, bunches of them!' Knowing mom as I do, that didn't surprise me." Before participating in the Marine invasion of Morningside, Dorothy had quite a life, but not a quiet life. She liked it that way.

Opha Mae Johnson got the ball rolling for lady Marines on August 13, 1918 when she enlisted during WWI for clerical responsibilities. During WWII, 23,000 women served as lady Marines in a wide variety of assignments. By the end of hostilities, 85% of all enlisted USMC personnel assigned to headquarters were women. Dorothy Turner became one of those very few, the proud, the lady Marines of WWII.

Raised in the Land of Lincoln, Dorothy mastered numerous male-dominated skills as a young woman, like plumbing, welding, and riveting. She said, "My father was a teamster truck driver back in the good old days. I danced many a dance with teamster boss Jimmy Hoffa, but every time I went 'juking' with friends, my dad would find out. It never failed, I'd be spotted by a trucker in a nightclub and he'd go tattletale to my father! It just didn't seem fair."

Dorothy asked her father's permission to join the Marines before the outbreak of WWII. She recalled, "The look on my father's face would have turned a person to salt, and he told me.... well, I can't really tell you what he said, but it meant 'No'!"

On the other hand, when she heard on a car radio that the Japanese had attacked Pearl Harbor, she joined the Marines anyway, without her father's approval. When asked what her father said when she did inform him of her enlistment, Dorothy replied, "Well, there you go again. To be honest, I can't really tell you what dad said, not without sounding too unlady-like."

Assigned to Camp Lejeune, Dorothy spent the war doing what she already knew how to do, plumbing, welding, and riveting. Dorothy said with a smile, "Rosie the Riveter was a pushover; the hard-hitting lady in the crowd was Dorothy the Welder." 'Hard-hitting' aptly defines Dorothy: Loyal, tenacious, and Leatherneck proud, she nonetheless dodged a court-martial and 'hard-hitting' was the culprit. Dorothy explained, "I was sleeping like a baby one morning when the barracks sergeant came in and whacked me on my foot in an attempt to wake me up. She made a huge mistake. I came up swinging and knocked her out cold." Albeit, the court-martial never materialized since Dorothy's enlistment records clearly indicated, 'hates to be rudely awakened.'

Dan, her son, confirmed the peculiarity. "I learned at an early age how to wake up Mom, tap her on the arm and run like hell!"

Dorothy met her future husband, Bud, on a blind date. She said, "Bud was a Navy Seabee during WWII. They were among the first Americans on Guadalcanal, which enticed Bud to put a sign on the beach that said, 'Welcome, Marines,' but he said the Marines failed to see the humor."

A young couple in love, Bud and Dorothy Turner

The Aleutians also required Bud's Seabee skills. While on the Aleutians, he met a young soldier from Georgia named Charles West, son of George West, owner of West Lumber Company and the First Federal Savings and Loans. Dorothy said, "Now you know how we ended up in Georgia. We came south after the war to help George build homes in Atlanta until my husband branched out on his own."

A mother of four boys and one girl, Dorothy became a Gold Star Mother after her son, Marine Pfc. Michael Barry Turner, was killed in action. He was 19 years old. In his honor, Bud and Dorothy established the Michael Barry Turner Memorial Scholarship program in 1969. For ten years, the Turners awarded three scholarships each year in sports, academics, and music to Towers High School in Decatur, GA. Bud coached the Midway Mighty Mites football team to mentor and help young men get off on the right foot in life.

For over 60 years, Dorothy purchased and distributed gifts to disadvantaged children during the Marine's Christmas Event Toys for Tots. If she needed assistance with loading or distribution, Dorothy drove to a local Marine Recruiting Center, marched in, then barked, "You, you, and you, get out here right now! I need your help." The recruiters never argued with Dorothy.

Dan attempted to relate his father's entrepreneur endeavors, "When I was just a kid, Dad owned a couple of nightclubs around Atlanta. If the piano player..."

Dorothy instantly interrupted, "They weren't nightclubs; they were dang strip joints! I never did like those hoochie-coochie clubs."

Dan cleared his throat. "Well, at any rate, if the piano player didn't show up, Dad would come get me and we'd sneak off to the club. I'd sit behind the upright piano, sort of out of sight, and played music all night long with some of the best jazz players in Atlanta. I was only nine years old when I started and didn't get caught by the police until I was twelve. Mom never knew about it or she would have killed both of us."

When I asked Dorothy if she ever found out her adolescent son was playing music in strip joints, she replied, "Yeah, Dan finally confessed when he was old enough to outrun me."

If any of her sons and/or their friends 'acted out' or showed a lack of respect for their teachers at Glenhaven Elementary School, Dorothy would receive a call from the principal. Basically, the man was calling in Marine reinforcements to administer disciplinary action. Dan clarified, "Mom would march to school and haul an unruly son and his friend out into the hallway and blister their unruly butts. Nobody complained back then, no lawsuits, no child abuse; nothing but good old-fashioned punishment from a concerned and tenacious parent. Plus, nobody wanted to mess with mom."

Dan recalled the numerous times he was in grocery stores with his mother when a rowdy kid was in the checkout line. "Mom would say, 'Lady, either get that child under control or I will!' The disorderly kid would take one look at Mom then settle down, like immediately."

When asked to tell a secret or two about Lady Marines, Dorothy stated, "Well, I can't do that, because if I told you they wouldn't be secret anymore, now would they?"

September 9, 2011: The Marine Corps League Devil Dogs Detachment #952 from Duluth, GA and the General Ray Davis Marine Corps League Detachment #1188 from Monroe, GA, plus over a hundred friends, family, community leaders, and Morningside residents joined in a celebration of life for Marine Dorothy Turner. Included in the event was a replica presentation and reenactment of the flag-raising on Iwo Jima.

A young lady at the celebration asked Dorothy, "So, Mrs. Turner, you were a Marine in WWII?" Dorothy cut her eyes at the woman and replied, "No, young lady, I *am* a Marine!"

On Oct 22, 2011, Marine Dorothy Turner departed this life and immediately marched up to and through the entrance to the Pearly Gates. The thunderclap created when the gates slammed against the backstops was heard worldwide. Saint Peter said nary a word.

THE LAST RAIDER

WWII Days at Falcon Field lived up to expectations once again this year. A gorgeous B-25 was on display to commemorate the 75th anniversary of the Doolittle Raid on Japan early in WWII. Reenactors, vendors, even operable B-25 gun turrets for the young and 'very thin' young-at-heart enthusiasts were part of the program. The B-25, a P-51, and a Dauntless Dive Bomber plus various military trainers also offered hops on the historic planes.

In the vendor hangar, the only remaining Doolittle Raider, 102-year-old Lt. Col. Richard "Dick" Cole, enthusiastically signed copies of his book and posed for photos with an admiring throng. I'd briefly talked with Lt Col. Cole at another event then penned an article on the famous Doolittle Raid, but this time a journalistic miracle unfolded. Cindy Cole Chal, the attractive, attentive and justifiably protective daughter of Lt. Col. Cole, recognized me from our first meeting and asked if I would like to join Sue Verhoef from the Atlanta History Center for Sue's scheduled interview with her father Sunday morning.

LT. Col. Cole, and Author Pete Mecca

I jumped at the opportunity. I've known Sue Verhoef for several years and didn't want to encroach on her interview turf unless she agreed. Her Atlanta History Center display was set up near our Atlanta WWII Round Table exhibit so I made a beeline for Sue's table. She happily agreed.

Sue led the interview, it was her show; as a result, my questions for Lt. Col. Cole were limited but based on a thorough knowledge of the Doolittle Raid. The sole survivor's physical health at 102 is still good yet waning, but his mental sharpness, eagerness, and wit is firmly intact. In short, Lt. Col. Cole was a real hoot! The questions and answers are forthright with insignificant editing for clarification.

THE INTERVIEW

Sue: What did you have for breakfast, Colonel?

Cole: Let's see if I can remember. Oh, eggs and biscuits.

Sue: Did you sell all your books yesterday?

Cole: Yes, I did.

Sue: Thank you for coming, Colonel.

Cole: Thank you for having me.

Sue: May we have your full name?

Cole: Richard E. "Dick" Cole. I was born in Dayton, Ohio on September 7, 1915.

Sue: Tell us about your background, your childhood.

Cole: Well, I was blessed to be born into a wholesome family arrangement, number five of the children. I grew up with good schools and a great high school. I would have to admit it was a happy environment. I always thought of myself as a happy child.

Sue: You were interested in aviation before the war. Why?

Cole: When I was old enough to ride through the neighborhood on my bike by myself, I guess about 10 years old, I'd ride down to the Army airfield near Dayton. I would sit on the river levy and watch what's going on, even without realizing who was performing on a particular day.

Pete: By 'performing' do you mean the barnstormers?

Cole: Yes, and it was all very interesting. There was an annual airshow in October, and I always went to that. Anyway, I made up my mind to either be an Air Corps pilot or a forest ranger.

Sue: So, how did everything unfold?

Cole: Well, I tried to map out a plan on how to get there, to be a pilot; but later on, I realized that wasn't much of a plan. The main reason I wanted to be an Army pilot, once a month Wright Field sent a Curtiss Hawk to hover over an Army truck, then it would quickly bank into the base. It was payday! That sounded good to me, plus I wanted to go after the bad guys.

Sue: What made you decide to join the military?

Cole: Between middle school and high school they offered an aviation course. I attended that, was 3rd out of 33 grads. For those interested, they gave us a chance to drop out of high school then attend another course. Quite a few of us did. The course was more toward aircraft mechanics, tearing apart a Liberty engine, there were nine of us, and we had to put it back together. The teacher had rigged a system to crank the propeller to see if the ignition system was installed correctly but we couldn't really start it.

Sue: And after that?

Cole: I spent a lot of time looking for a job! People today don't realize how hard it was to find a job back then. Anyway, I found a worthwhile job, $75.00 a month plus room and board. I did that for a couple of years, then started college. Before college I went to Wright Field for training in the Army, I passed all the tests, but in the end they turned me down. See, the Air Corps was under the Signal Corps, and they didn't take anyone in need of medical attention. I had two bad teeth and a set of tonsils, so I was turned down. After that, I figured I should go on to college and learn all I could about Smokey the Bear.

Sue: So how did you get into aviation?

Cole: I started college, but between my sophomore and junior year I got the word that Uncle Sam had this program, he would pay for you to get a pilot's license. But if you qualified and received your license, you had to make yourself available for any kind of emergency.

Sue: So, that's how it started?

Cole: Yeah, that's how it all got started. I was working during the summer in my job and went to…wait, let's back up a minute. I got my pilot's license during September of '41 and…

Pete: On what type of aircraft did you train?

Cole: A Taylorcraft. So, I graduated then they told me to wait. Well, by November I got tired of waiting and went to enlist.

Sue: How did your family feel about you enlisting?

Cole: They were glad to get rid of me. So, no objections from family members. I went to Fort Thomas for a month then got orders for St. Louis. I was lucky enough to get through that program then ended up at Randolph Air Force Base. After graduation, a few of us were sent to Pendleton, Oregon to join the 17th Bomber Group. We trained here and there and ended up with a fairly good bomb group.

Sue: How did you find out about a secret mission needing volunteers?

Cole: Well, at Pendleton we were all copilots and I was crewed up with a gentleman that had quite a lot of time in B-25s, that's how we welded into a squadron. When I arrived at Pendleton the group made a move to Spokane, Washington. We spent the summer there, then got orders for Columbia, South Carolina. We trained with the Army Air Corps plus ground forces. On December 6, 1941, we were at March Field on our way back to Pendleton. Of course, the next morning, on December 7, 1941, the war started. Back at Pendleton, they changed our bomb racks to accommodate submarine bombing. We were on submarine patrol along the coasts of Oregon and Washington until February of'42. Then we got orders to return to Columbia. We were at a newer airbase, it wasn't quite finished, mostly mud and water. We lived in tents. Each squadron had an information board that we were required to examine daily. One day we read they needed pilots for a secret mission. Well, our instructor was a great guy, a really good pilot, and he put his name on the list. I thought that was a pretty good idea, so I put my name on the list, even if I didn't think it was such a good idea after I did it.

Sue: Why were you chosen for the mission?

Cole: They chose the pilots, don't know how they did it, but we started training for the mission. Well, the instructor I trained with became ill, so I needed a pilot to fly with. I went to the Ops Room asking questions and they said, 'you still want to go?', and I said, 'Yeah,' so they told me, 'the old man is coming in this afternoon. He'll take care of it.' Anyway, Jimmy Doolittle showed up about an hour after that. They told him I'd been chosen as his copilot, he said, 'Fine', and that was that.

Pete: Did you know Doolittle was 'the old man', your leader?

Cole: Yes. I was fortunate to know who he was and what kind of pilot he was because I'd seen Doolittle perform at airshows.

Sue: What do you remember about your training at Eglin Field, what stands out?

Cole: Doolittle wanted us to have 50 hours of training, but with launch scheduled for April 19, he only had 90 days to get us ready. We ended up with about 30 hours of training. Turned out that was enough to do what we needed to do, and we did.

Sue: Tell us about your trip to the USS Hornet, the carrier to carry you to your destination. Did you watch them lift the B-25s onto the deck?

Cole: We were in San Francisco. The Navy has that kind of equipment, big cranes, everything was hooked up and went well.

Sue: Tell us about the crossing. Were you ever on a ship before and did you get seasick?

Cole: No, I'd never been on a ship before, and no, I didn't get seasick.

Sue: Did you know you were going to take off from a carrier?

Cole: Well, it was pretty obvious we would be taking off from a carrier, but we didn't figure Japan as the target. We figured we'd take off and land on some island and get on with fighting the war. But two days out, we were told our target was Japan. There was a lot of jubilation, a lot jumping up and down and hollering, but after that things got real quiet, that too pretty obvious as to why. Doolittle said if anyone wants to get out, to back out, there would be no repercussions. Nobody backed out.

Sue: Tell us about the launch, when you got the word to 'go'.

Cole: By that time, we had a pretty good chunk out of our brains due to all the training, so we truthfully didn't think it posed a big problem. We had navigation, classroom instruction, so with everything considered, we had the training to do the job.

Sue: What was it like to launch? Was it a scary moment?

Cole: Surprisingly enough, with our training we were okay. The only concern was the weather, but that turned out okay. We figured the launch would be challenging, so we positioned a B-25 in the middle of the deck which gave us six feet of clearance on both sides. We painted a white line on our left gear and one for the nose wheel so we could steer down those lines and it worked okay. We didn't consider the launch a big problem.

Pete: Would a night launch have been considered?

Cole: Yes, Doolittle had considered a night launch possible. He approached the Navy with the idea, but they wouldn't buy it. They feared one of the bombers would drift over and hit the island (command structure) and put the aircraft carrier out of commission. We only had four carriers in the Pacific at the time and the Navy couldn't afford to lose one.

Sue: What was the mood of the crew during the flight to Japan?

Cole: We didn't get keyed up or anything, we didn't have any Japanese aircraft try to intercept us. When we got to Japan it was like flying into Miami Beach. People were playing on the beach and working on their boats, all kinds of activity. And we hadn't made any mistakes, so we continued on.

Sue: Then you were en route to China?

Cole: Well, once we dropped our bombs we were, but it wouldn't have done any good to have a base to land on in China. The entire area was socked in, couldn't see a thing, and we didn't know where we were.

Sue: When did you get the orders to bail out?

Cole: Well, the whole flight took 13 hours…four to Japan, nine hours across the water to China. No particular or dramatic things happened. We ended up thinking about what could happen, especially after Hank, our navigator, handed me a note saying we were going to end up about 180 miles short of China. We didn't know what to think about that. But we got to China with fuel to spare, a tailwind helped us.

Pete: No homing beacon?

Cole: No. An aircraft with a homing station was supposed to basically 'greet' us into China, but their plane crashed en route and killed all the occupants so we lost the radio equipment. We didn't think of it at the time of launch, but as we circled the carrier and looked down at the 'heads-up' board for our course to China, we knew we had separated ourselves from civilization. In our zeal to make the bomber as light as possible to take on more fuel, we tossed out all the long-range radio equipment. The only communications we had was a command set, a range of about 45 miles. Had there been an emergency, sixteen B-25s would have been floating around in the Pacific.

Sue: Had you trained for a bailout? Had you used a parachute before?

Cole: No. And not one since.

Pete: Colonel, what did you think of the B-25?

Cole: The only thing I can say about that is, all through training I drove around in a Model A Ford. Going from a Model A Ford into a Mitchell B-25 bomber is like crawling into a Mustang. Great aircraft.

Pete: What was your altitude en route to Japan.

Cole: About 200 feet above the water. We stayed at 200 feet, except climbing to 1500 feet to bomb and later to bail out. Other than that, we stayed at 200 feet.

Pete: What was your target?

Cole: Our primary target was the northwest section of Tokyo because we carried incendiaries so we could light up Tokyo for the rest of the bombers, like a reference point.

Pete: Did you dodge any flak or enemy fighters?

Cole: No fighters, but pulling off the bombing run we had flak coming up at us…bounced us around a bit.

Pete: After you bailed out, you landed in a tree, correct?

Cole: Well, that was the best landing in that situation. The terrain was very mountainous, so I was pretty lucky and waited until morning to cut myself loose. Just a few scratches. Although I found out later the Army doesn't hand out Purple Hearts for self-inflected injuries. I was luckier than Doolittle though, he landed neck-deep in a pile of manure.

Pete: Your entire crew made it back, correct?

Cole: The whole crew survived. Some of the Raiders returned to the States, but 26 of us stayed. Six of us went to the CBI Theatre (China-India-Burma). I got checked out on a C-47 and flew them for about 14 months, including the Hump (flying across the Himalayan Mountains).

Sue: Your last comments, sir?

Cole: Well, when I think about it, the mission was not a highly dangerous affair. You could do something about it if there was a problem. But, looking back, I'd say we were pretty lucky.

Lt. Potter, navigator; Lt. Col. Doolittle, pilot; Sgt. Braemer, bombardier; Lt. Cole, copilot; SSGT Leonard, flight engineer & gunner

When Lt. Col. Cole found out I had served in Vietnam, he said, "There's another time I was lucky." I asked, 'So, you were in Nam, too?' Cole replied, "No, I was lucky that I didn't have to go!"

Lt. Col. Cole flew 21 different aircraft during his career, the last one a T-33 Shooting Star jet fighter. Of the 80 Doolittle Raiders, 69 survived the mission. Three were killed in action. Eight were captured: three executed by the Japanese, one died in captivity, the other four returned home after the war. Of the 28 Raiders who eventually flew in the China-Burma-India Theatre of Operations, five were lost. Nineteen former Raiders flew in the Mediterranean Theatre and four of them were killed. Nine

Raiders flew in Europe, one was lost. Of the 63 Doolittle Raiders who survived the war, 102 year old Lt. Colonel Richard "Dick" Cole is the sole survivor.

My heartfelt thanks to Cindy Cole Chal and Sue Verhoef for the opportunity to participate in the interview of a lifetime.

"I have been luckier than the law of averages should allow. I could never be so lucky again."

—Jimmy Doolittle

POLITICAL SCAPEGOAT

"And Aaron shall cast lots upon two goats; one lot for the Lord, and the other lot for the scapegoat."

—Leviticus 16:8

In biblical times one goat was sacrificed and a second goat, the 'scapegoat', was exiled into the wilderness as an atonement to 're-move the burden of sin' for others. In modern times, the 'scapegoat' is a person unfairly blamed for the shortcomings of others.

On December 7, 1941 at approximately 7:48 am, the first of two waves of Japanese aircraft swept in from northwest of the Hawaiian Islands to unceremoniously welcome the United States of America into WWII. It was broadcasted as a 'sneak attack', or a 'surprise attack', and in President Roosevelt's famous words the

Japanese attack was destined to become, 'a date which will live in infamy.'

In approximately two hours, 21 warships (including eight battleships) had been sunk or damaged. Of the available military aircraft, 188 were destroyed and another 159 damaged. American casualties: 2,403 killed and 1,178 wounded. As Japan continued to run wild throughout the vastness of the Pacific Ocean and Southeast Asia, the American press, its people, and the government demanded to know why and how the Pearl Harbor tragedy unfolded. The scope of the catastrophe was visible, but to many critics the unpreparedness of American military power in Hawaii bordered on criminal. Heads were about to roll and stones were about to be tossed.

When protecting an adulteress from executioners, Jesus said, "Let him among you who is without sin be the first to throw a stone...." In 1941, Washington, DC was a political caldron of sinners with no qualms about tossing stones to protect their own careers and hindquarters after the attack on Pearl Harbor. Two men were targeted for dereliction of duty, demoted and sacked from their responsibilities, and both officers were denied a court martial each demanded to properly defend themselves: Admiral Husband Kimmel, our Commander in Chief of the Pacific Fleet at Pearl Harbor, and General Walter Short, Army Commander on Oahu. Both commanders shoulder some of the culpability, but they were not the only Pacific commanders caught with their khakis down. General Douglas MacArthur in the Philippines was notified of the sneak attack on Pearl Harbor seven hours before the Japanese caught his aircraft on the ground. MacArthur lost the Philippines, had over 70,000 men killed or taken prisoners, yet was awarded the Medal of Honor and went on to become a national hero.

Admiral Husband Kimmel

Enter Tom Kimmel. As Commander of the Atlanta World War II Round Table, I had the honor to introduce Tom as our guest speaker on June 18, 2015. Tom is a noted Pearl Harbor scholar, a former FBI agent, and eldest grandson of our Commander at Pearl Harbor, Admiral Husband Kimmel.

Tom Kimmel

Tom's credentials are impeccable: A graduate of the US Naval Academy, Tom served on three warships during the Vietnam War and attended John Marshall Law School before joining the FBI in 1973. For more than 25 years, special agent Kimmel investigated organized crime in Cleveland, OH, worked with the House Appropriations Committee Investigation and Surveys at CIA Headquarters and supervised the FBI in East Texas. He headed the Labor Racketeering Unit at FBI HQ and the National Drug Intelligence Center in Johnstown, PA; served on the President's Council on Integrity and Efficiency and as the Assistant Agent in Charge of the Philadelphia FBI Division supervising the Foreign Counter-Intelligence and Terrorism Programs during the first bombing of the World Trade Center in 1993.

After retirement, Tom served as an FBI consultant addressing major spy scandals in the FBI and CIA. He has appeared on *60 Minutes, Discovery Channel, National Geographic Channel, War Stories with Oliver North,* and the *National Press Club* in Washington, DC.

Tom tendered an outstanding presentation in defense of his grandfather, plus provided enough food for thought to nurture the 7th Fleet. Convincing and concise, after hearing Tom's lecture, one could easily join the ranks of Admiral Kimmel's defenders and supporters. Albeit, to present this story I dug deeper, real deep, using numerous research engines and transcripts, read in detail five Pearl Harbor classics of conflicting hypotheses, plus scrutinized all 14 of the notorious 14 coded transmissions from Tokyo to its embassy in Washington from Dec 6 to Dec 7, 1941.

The term 'military intelligence' has offhandedly been referred to as an oxymoron. Having served in Air Force Intelligence, I have on occasion agreed with that statement. But if one analyzes the calamity of Pearl Harbor, it must be remembered that in 1941 our civilian and military leaders did not have spy satellites, GPS, supersonic SR-71 spy planes, up-to-the-minute cable news, cell phones, or a laptops armed with Google. During my first class in Intelligence school we were told 'don't believe anything you read and only half of what you see.' No truer words have ever been spoken, especially when one considers the devious spin-doctors infesting today's political climate and a biased mainstream media.

Pearl Harbor, Dec 7, 1941

Nonetheless, dive into the perplexity of eight formal investigations during WWII (the last investigation a joint congressional analysis beginning in November of 1945) and you'll take a roller coaster ride of accusations, slurs, favoritisms, biased reports, the lack of common sense, recommendations based on limited knowledge, and even worse...the unlawful destruction of military records.

He-said-she-said WikiLeaks journalism will not contaminate this article. The irrefutable hard facts will be divulged so the reader can derive his or her personal conclusions, pro or con, as to why only two officers bore the brunt of blame for Pearl Harbor.

In early 1941, Adolf Hitler was preoccupied with England and the Soviet Union. No credible evidence exists of immediate plans for Germany to instigate a war with America by striking its overseas possessions or the American heartland. The direct threat lay in the Pacific: Japan. Aware of Japan's aggressive strategies, the President ordered the Pacific Fleet to move from San Diego to Pearl Harbor in the spring of 1941.

Pearl Harbor: the massive American fleet sat poised to intervene in the Pacific. To what extent was the fleet protected? A brief list.

1. Four SCR-270 radar units were set up around Oahu. Limited spare parts, poorly trained personnel, the units could not identify friend from foe, and only manned between 0400 & 0700.
2. Out-of-date American fighter aircraft. The Japanese Zero flew circles around our P-26, P-36, and P-40 fighter planes. Pre-Pearl Harbor descriptions, observations, and warnings of the Zero's capabilities were deemed inaccurate because it was believed a fighter plane could not do what the Zero was reported capable of doing.
3. The need for long range patrol aircraft. Admiral Kimmel had 48 long range patrol aircraft, seven of which were un-

der repair. To cover the area effectively, Kimmel needed about 250.

4. Pearl Harbor's shallow water was considered an adequate deterrent against torpedo attack. The insightful Japanese simply developed a torpedo to travel in shallow water.
5. The firepower from fleet guns and Army anti-aircraft guns would successfully protect Pearl Harbor. Wishful thinking, at best, since the ammunition was kept under lock and key and unavailable.
6. General Short believed the Navy was there to protect his assets; Admiral Kimmel believed the Army was there to protect the fleet.
7. The seas north of the Hawaiian Islands were considered too rough, especially in winter, for a naval assault force to safely navigate. The Japanese gambled; the Japanese won. Other, more frequented sea lanes were the predicted route of any Japanese aggression, without consideration on how a huge fleet could steam the normal sea lanes unnoticed.

The above list is incomplete. Dozens of accusations, some valid, some manufactured, were hurled at Kimmel and Short. Yet most of the significant allegations against these men are trivial and pointless when undisputable evidence revealed that neither commander had access to the following information:

In late November and early December, 1941, a Japanese spy at Pearl Harbor was ordered by Tokyo to send details on American warships at Pearl, their moorings, the depth of the harbor, departures, arrivals, barrage balloons, anti-aircraft positions, number of aircraft stationed in Hawaii, and any significant information on the defenses at Pearl. This is the type intelligence required by an adversary to preplan an attack; Admiral Kimmel and General Short would have recognized this information as such. The spy's phone had been tapped, all the information recorded, and passed on to Washington. NONE of the spy's activities were reported to Short or Kimmel.

The most treacherous oversight bordered on reckless negligence by the Washington hierarchy: the Japanese code had been broken by military intelligence, the secret program dubbed MAGIC. MAGIC was the most important top secret intelligence program of the Pacific War. The MAGIC decoding machine was even made available to the British, vis a vis Winston Churchill. So secret was MAGIC that high ranking personnel, including General Marshall, lied under oath to protect the program. And there sat Admiral Husband Kimmel, Commander of the vital Pacific Fleet at Pearl Harbor, entrusted to defend and protect his country, his sailors, and his vessels. Kimmel's Pacific Fleet was the only real barrier to halt or restrain the Japanese, an enemy known to be planning an attack somewhere, at some time, and soon.

Admiral Kimmel WAS NOT given a MAGIC decoding machine, nor even informed of its existence. He was politically and militarily blindfolded with his hands tied behind his back. Intelligence is only good when shared with the people that need it most. To argue that the Commander of our Pacific Fleet at Pearl Harbor did not have 'a need to know' is ludicrous, if not felonious.

The 1945 Joint Congressional Committee Investigation, after reviewing numerous MAGIC documents released by the Truman Administration, came to a final conclusion on culpability: Blame should be allotted to all the principals, which included the two commanders at Pearl and the Navy and Army War Departments. It also concluded that Secretary of War Henry Stimson, Secretary of the Navy Frank Knox, the Chief of Naval Operations Admiral Harold Stark, Chief of Staff General George C. Marshall, and Chief of the War Plans Division Brigadier General Leonard T. Gerow, to be culpable for the disaster.

An even longer list of minor players did not bode well for the military nor our politicians in Washington, DC. Admiral Husband Kimmel and General Walter Short paid the price for being at the wrong place at the wrong time in the service of their country. Scapegoats? You decide.

One of Admiral Kimmel's lawyers wrote to him in 1953, "Pearl Harbor never dies, and no living person has seen the end of it." However, on May 25, 1999 the United States Senate approved a resolution stating that, 'Kimmel and Short had performed their duties competently and professionally', and added that, 'our losses at Pearl Harbor were, 'not the result of a dereliction of duty.'

Perhaps Senator Strom Thurmond of South Carolina said it best, "Kimmel and Short are the final two victims of Pearl Harbor."

JACK SIMPSON, SURVIVOR OF ANZIO

Jack in basic training

Depression era movies featuring the derring-do adventures of G-men (F.B.I.) collaring thugs with names like Baby Face Nelson, Pretty Boy Floyd, and John Dillinger inspired a young man from Pennsylvania to submit an application after high school graduation for fingerprint classification training. Accepted for training,

Jack Simpson worked with the F.B.I. for eight months until the draft board classified him 1-A. "I knew I was a goner," Simpson stated. "So I joined rather than accept the draft."

Following in the footsteps of an uncle, David Bougher, a recipient of the Silver Star for gallantry during the Japanese attack on Pearl Harbor, Simpson joined the Army. After basic training at Camp Butner, NC, he was given specialized training on the track-mounted 105mm howitzer. Simpson said, "I trained on a 50 cal. machine gun, was the weapons loader, plus I set the elevation. After they deemed us qualified, we shipped out with the 45th Infantry Division, 179th Infantry Cannon Company."

After brief fighting in North Africa, Simpson boarded the British vessel HMS *Derbyshire* for Naples, Italy. He said, "Once in Naples we began the staging for an amphibious landing behind German lines." That amphibious landing would enter military history books as *The Battle of Anzio.* The nearby seaside resort town of Nettuno was also part of the operation but not as well-known nor given the limelight afforded the beaches of Anzio.

The Anglo-American invasion forces appeared formidable: the U.S. 3rd Infantry Division, the 504th Parachute Infantry Regiment and 509th Parachute Infantry Battalion, the 751st Tank Battalion, three battalions of Army Rangers, and the 45th Infantry Division. British forces included the 46th Royal Tank Regiment, two Commando Battalions, and the British 1st Infantry Division.

Support for the ground forces included at least 2,600 Allied aircraft and three Naval Task Forces of approximately 375 combat, landing, and supply ships. A diversionary attempt sent other naval units 40 miles north of Anzio to bombard the coastal city of Civitavecchia. But men and material are useless in war if one indispensable component is lacking as it was on the beaches at Anzio and Nettuno: Initiative.

At first optimistic of their chances, skepticism wormed its way into the military judgment of seasoned commanders like Generals Clark, Truscott, and Lucas. The commander of the invasion, Major General John Lucas, received a two-prong objective: '*to divert enemy strength from the south and, in anticipation of a swift and violent enemy reaction, to prepare defensive positions.*' When a military unit is issued an order to '*prepare defensive positions*' it sends a smoke signal to an opposing force: Come and get us.

Notwithstanding, at 0200 on January 22, 1944, Allied forces storming the beaches of Anzio and Nettuno achieved one of the most stunning military achievements of the war: They caught the Germans sleeping, completely off guard. Army Rangers occupied Anzio as the 509th Parachute Infantry Battalion secured Nettuno. Other objectives were attained before noon and by that evening the 36th Engineer Combat Regiment had cleared the mine fields, put down corduroy roads, bulldozed exits from the beaches, and prepared the port of Anzio to receive supply ships. By midnight, over 36,000 Allied troops with their equipment in hand crowded the beaches.

In just a few days, the Allied forces had moved over seven miles inland. German resistance was weak at first, yet gained in vigor near the towns of Campoleone, Cisterna, and Littoria. Now the fog of war reared its ugly head. General Truscott was ordered by the corps commander to halt the offensive and to 'consolidate and reorganize' the beachhead. On the opposing side, the Germans simply could not believe their good luck.

Anzio Annie - captured

Crack German units like the *Hermann Goering Divisions* and the *4th Parachute* streamed into the region. Hitler ordered experienced combat units from Germany, France, and Yugoslavia to reinforce factions of the 3rd *Panzer Grenadiers* and *71st Infantry Divisions* already moving into the breach at Anzio beach. Hitler called the invasion '*The Anzio Abscess.*'

American General Lucas launched a two-pronged offensive against the Germans on January 30. Unknown to Lucas and Allied commanders, their attack was heading straight into a marshalling center for 36 enemy battalions preparing their own counterattack on February 1. In one incident, 767 U.S. Army Rangers marched into the claws of the *715th Motorized Infantry Division* soon strengthened by armored units from the *Hermann Goering Division.* The Rangers were exterminated; only six made it back to Allied lines.

Predictably, the Allied offensive ground to a halt. Generals Clark and Lucas ordered Truscott to 'dig in', then sent in Allied reinforcements, including supplemental antiaircraft and artillery groups. The crowded beaches of Anzio and Nettuno soon housed more than 100,000 Allied warriors. The beachhead resembled the trench-warfare stalemate of World War One. Filthy rain-soaked dugouts, trenches, and foxholes honeycombed the oceanfront as Allied soldiers hunkered-down to take the best the Germans could throw at them. And throw the Germans did.

Two enormous railroad guns, *Leopold* and *Robert,* pounded allied soldiers mercilessly. Initially thought to be only one railroad gun, the trapped soldiers on the beach called these monsters *Anzio Annie.* Tucked safe and sound behind the Alban Hills more than 20 miles from the beaches, the two guns could hurl 15 rounds per hour at targets up to 40 miles away. To operate properly, each gun required a crew of 42 men and officers.

The battle stalemated, waning from offensives and counter-offensives; attacks and counter-attacks, snipers, strafing, misery, maiming, destruction and death, and disillusionment for the better part of four months. Eventually, one element the Allies had that the Germans did not was resources. No military force can last indefinitely fighting hunger and short supplies against an unwavering well-fed and well-supplied enemy.

Jack Simpson and his 105mm tracked howitzer crew were among the first men ashore on Anzio. Simpson recalled, "We dug in immediately. Our foxholes were right next to the 105 and we covered up with anything we could get our hands on. The Germans shelled us, their planes strafed our position, snipers tried to pick us off, and spent anti-aircraft rounds from the Navy ships peppered our dugouts. There were several ways for a soldier to die on Anzio Beach."

The night offered no reprieve. Simpson stated, "Parachute flares turned the night into day, so the beach was well-lit. Our Navy pounded the Germans all night, then the Germans pounded right back. Sometimes we'd fire 10 rounds from the 105 every 30 minutes just to keep the Germans on their toes. Of course they would shoot right back to keep us on our toes, too."

Death stalked the beaches, the sea, and the sky. Simpson recalled, "Men would disappear in an explosion, so would a ship. We'd watch big formations of B-17s flying overhead, saw them get hit, catch on fire, then spiral out of control. The most sickening thing we saw was the German fighters swooping down to machine gun our helpless airmen in their parachutes."

On his battlefront experience dodging the deadly firepower of *Anzio Annie,* Simpson said, "Just imagine a freight train passing right over your head; that's exactly what it sounded like."

After four months of stalemate slaughter, the Allied 'breakout' finally occurred in late May of 1944. The butcher's bill: around 29,200 allied combat casualties and over 37,000 noncombat casualties. The Germans fared slightly better with 27,500 combat casualties.

Eric Fletcher Waters, father to Pink Floyd's bassist and songwriter Roger Waters, lost his life on Anzio Beach. Pink Floyd's larger-than-life song *When the Tigers Broke Free* recounts the actions at Anzio and Nettuno. Other notable participants include America's most decorated soldier of WWII, Audie Murphy, plus future star of the long-running TV western 'Gunsmoke', James Arness, forever remembered as Sheriff Matt Dillon. Arness walked with a lifelong limp due to the severe wounds he received on Anzio.

Simpson's unit, the 45th, was first to arrive in Rome. "The Italians showered us with flowers, excitement, and a whole lot of wine," he recalled with a big grin. But Simpson soon left one hotbed of activity to enter another, landing at St. Tropez in Southern France during August of 1944. On his experience with the French, he stated, "They were happy to see us, too. We were offered food, wine, hand-shakes, and an occasional kiss."

Preoccupied with the Normandy Invasion, Germans in the southern France area finally made a stand near Epinal in the Rhine Valley. Stalled briefly while engineers completed a pontoon bridge, Simpson decided to take a hike to a nearby hill and take a few pot-shots at rabbits. "They were huge," he said. "The size of baby kangaroos. I wasn't trying to kill the rabbits, just spook them, and watch them jump." Unexpectedly, 20 German soldiers stood up from behind a wall with their hands in the air. Simpson recalled, "They figured we had them surrounded so they wanted to give up. One spoke perfect English and asked for food. They were starving." Simpson gave them what food he carried then marched the German POWs back to his unit. "My buddies were dumbfounded, but my 20 prisoners were nothing if compared to the 45ths two-year total of 124,000 POWs."

The 45th crossed the Rhine River into Germany and fought all the way to Munich by war's end. There, Simpson and the 45th joined the Army of Occupation. A recipient of two Bronze Stars for bravery, Simpson returned to college after the war to earn a Master's Degree at George Washington University in Washington, DC.

He returned to the F.B.I. as a special agent and played vital roles in numerous high-profile investigations, including the assassination of Dr. Martin Luther King, Jr. Simpson retired after 23 years of service, but a proverbial rocking chair did not appeal to a man accustomed to hard work and a desire for achievement. He served as a Superior Court bailiff for 20 years, worked as a county bailiff, and now at the young- at-heart age of 95 continues to apprehend the bad guys as an investigator for the Newton County Sheriff's Office in Georgia. Simpson authored three books, writes a weekly newspaper column, is a notable civic leader, and admired public speaker. His viewpoint is lively and to the point: "Stay healthy and active as long as possible. I plan to hang around for a long time."

SHORT STORY BUT A LONG LIFE

Chances are, an individual old enough to remember The Great War, now referred to as World War I, and bounced along dusty

Georgia back roads in a rattle-trap Tin Lizzie would possess stories few of us have heard. And chances are, a young lady who lived through The Great Depression and didn't set eyes on a parked 'flying machine' until attaining the age of 27 would be a wealth of knowledge concerning her life, the country, and a thousand other things she had witnessed leading up to the year 2011 when she reluctantly granted the first interview ever concerning her service in World War Two and the years in between. But chances are, you would be wrong.

Elizabeth "Bubba" McClain is an American icon, a well-bred Southern lady of The Greatest Generation who remembers when ladies acted like, well, ladies. Her niece, Mrs. Louise Melton, contacted me to ask if I'd be interested in interviewing her aunt, a retired veteran of WWII and resident of the Morningside assisted living facility in Conyers, Georgia. Mrs. Melton said her aunt served in WWII as a registered nurse, entering the Army as a 2nd Lieutenant in 1942 and retired as a Major in 1962. The decision to interview was not difficult. With Mrs. Melton acting as intermediary, we teamed up to interview a lady with deep-seated devotion to family and country, and as I was soon to discover, a woman of very few words.

Southern as buttered grits and cornbread, Elizabeth McClain could be anyone's grandma, or more true-to-life, their great-great grandma. Ambulatory only for short distances, she arrived via wheelchair for her interview in the Morningside dining hall. Petite with a pleasant face accompanied by a pleasant smile that never faded, she shook my hand and said, "Nice to meet you."

Now, here's the thing: 'Nice to meet you' was the longest sentence she produced all afternoon. Thankfully her niece, Mrs. Melton, filled in the areas sorely in need of dialogue. The lack of discourse was not due to senility or a disinterest for the interview, rather, of an acceptance that whatever happens between birth and your last day on this side of the grass is really no big deal. I suppose folks should politely accept her peculiar philosophy since Elizabeth grew up in the Land of Cotton, living her entire life with 'Bubba' for a nickname.

More clearly defined, much of Elizabeth's dialog penned in this story is actually clarifications expressed by her niece, Mrs. Melton, in an attempt to expand on her aunt's concise answers, such as: "I reckon so," or "It was okay" or "Sure was" or "I don't remember" or "Ya got me." My personal favorite response from Elizabeth was, "Is that important?" Major Bubba, as Elizabeth is called by her Morningside family, definitely required yours truly to improvise normal interviewing techniques.

ELIZABETH "BUBBA" MCCLAIN

When Dr. John W. McClain made house calls in Mitchell County, GA, his vivacious daughter Elizabeth would run to the barn so she could turn the starting crank handle on the family's Model T Ford. It was always a big thrill for young Elizabeth to drive the 'Tin Lizzie' from the barn to the front of their house. She said, "The other town doctor in Pelham wouldn't make a house call if he was playing a game of bridge, so my dad got the business." McClain's grandfather was also a doctor. She said, "I knew by the 5th grade I'd either be a nurse or a missionary; it's sort of the family thing to do."

During the Great Depression, her father received payments for his services with chickens and corn, an occasional hog, and on rare occasions, real money. McClain recalled, "We grew our own vegetables and my mother made lye soap in the backyard. We knew about the Great Depression but my family was lucky, we had food, clothes, and a roof over our heads." Asked to explain her nickname, she replied, "The little girl across the street had difficulty pronouncing words so she called my grandfather 'Doc', my father 'Doctor Me Doc' my mother 'McMomma', and she called me 'Bubba'. I really don't know why; I never chewed tobacco or drove a pickup truck."

The small township of Pelham was the largest municipality in thinly populated Mitchell County. The central gathering place in Pelham was the Hand Trading Company. McClain recalled, "It was a four-story building that served as a post office, grocery store, drug store, a funeral home, hardware story, clothing store, plus a few other things. One of my brothers delivered mail from Hand Trading in a horse and buggy."

After graduation from high school in 1929, McClain chose to remain in Pelham to study advanced courses in biology and chemistry while caring for ailing parents. Eventually her six siblings took over parental care, giving McClain the opportunity to move to Atlanta in 1939 to study nursing at the old Piedmont

Hospital on Capital Avenue. She was in her second year of studies when aviators of the Rising Sun paid an unexpected visit to Pearl Harbor.

She vividly recalled that date of infamy: "I was shopping in Davison's Department Store when they announced that the Japanese had bombed a place called Pearl Harbor. I don't remember what people were saying in the store but I knew it meant we were at war. Within a few days, an Army nurse spoke to our graduating class. I liked her presentation and what she had to say, so I joined up."

As a licensed Registered Nurse, McClain bypassed military basic training and was commissioned as a 2nd Lieutenant on September 5, 1942. "My first assignment was Camp Wheeler, GA," she said. "Many of the soldiers were hospitalized for heat related problems during training." McClain completed two years at Camp Wheeler before reporting for duty at Fort McPherson, GA for an additional three years. "I was at Fort Mac when the war ended," she said. "We treated combat veterans as well as boys hospitalized for other illnesses."

McClain remained in the Army after WWII. When asked why she chose to stay in the military, she said, "It seemed like the thing to do." Asked if she enjoyed the Army, she stated, "I'm not sorry I went in." When asked about Army food, she said, "Well, I ate it."

Her next port-of-call: The war-torn 'Pearl of the Orient', Manila in the Philippines. McClain said, "We left from San Francisco and sailed across the Pacific Ocean. It was a real long voyage and we were happy to see land. But the thing is, after docking in Manila, we were told that several of the nurses would remain in the Philippines and the rest would be sailing to Okinawa. One nurse pitched a temper tantrum about going to Okinawa, so I volunteered to take her place."

Major Bubba retiring in 1962

Okinawa proved to be a demoralizing 18-month assignment for McClain. "I did not like Okinawa and I did not like the typhoons," she adamantly stated. "One of those typhoons made landfall and wiped out half of our barracks. That was okay with me because I lived in the other half. But the darn thing reversed course the next day and hit Okinawa again; demolishing my half of the barracks. I hated Okinawa!"

Proceeding with the interview to hopefully leave Okinawa in our wake, McClain continued, "My next assignment was Fort Jay on Governor's Island, New York. I loved Fort Jay, it was great duty. I would have stayed for the remainder of my career if the Army had let me." When asked why Fort Jay was so ideal, she replied, "It wasn't on Okinawa!"

Future assignments included Hot Springs, AR, Fort Eustis, VA, and Frankfurt, Germany. Major Elizabeth "Bubba" McClain retired from the Army in December of 1962. She returned home for a short stay in Pelham to conduct family business then moved in with her three sisters in Atlanta. Together they nurtured two ailing brothers-in-law and, in due course, each other. Major Bubba never married.

During the interview, McClain said, "When I was at Camp Wheeler during the war, a young nurse came into the ward and noticed a black soldier in one of the beds. She said, 'What's he doing in here?' The soldier pointed at me and said, 'She knows.' I told that uncouth young nurse, 'Listen here, this man is my patient, he's an American soldier, and you can just vamoose!' Well, she skedaddled, and I was glad she did. What she said wasn't proper."

Major Bubba was born January 11, 1912. She'll be one hundred years old her next birthday.

Postscript: Elizabeth "Bubba" McClain slipped gently into the good night on October 31, 2011, 73 days short of her 100th birthday.

UNARMED AND OUT OF FUEL

Dawn presented the harbor with serene winds and a picturesque pastel blue sky. Sailors sprang from shipboard berths; soldiers stretched and yawned in Army bunks as their brother seadogs, aviators, grunts, and Marines returned from a weekend pass or shore leave before stumbling into their bunks and berths to sleep off a night on the town. All-night poker games were breaking up, the gamblers headed for breakfast in the nearest chow hall, while others still ashore awoke in strange places next to strange women they didn't recognize. Thousands of service personnel were already up, dressed, and raising the colors on Navy ships and Army airfields. Such was the laidback lifestyle around and in Pearl Harbor, Hawaii on December 7, 1941.

The day before, on December 6, 1941, a flight of twelve B-17 Flying Fortress heavy bombers took off at fifteen minute intervals from Hamilton Field near San Francisco. Their destination: Hawaii, a long and tedious 15-hour journey before continuing on to the Philippines under the code name: PLUM. The pilots and copilots complained about steel protective plates hastily welded to the backs of their seats before takeoff. The metal plates were uncomfortable, tough on the backbone, and in the opinions of the flyboys a silly waste of Uncle Sam's money. The tough steel would later save their lives.

A young co-pilot on one of the B-17s, 2nd Lieutenant Ernest Leroy "Roy" Reid recalled their arrival. "It was around 0800. We were on a long base leg approach to Hickam Field when I noticed dense black smoke curling above the harbor. Our pilot, Captain Swenson, had flown to Pearl Harbor before so I asked the Captain about the smoke. He replied, 'Oh, don't worry about it. That's just local natives burning off sugarcane.' I thought his answer odd and I kept thinking, 'How do you grow sugarcane on top of water?'"

Reid's B-17

Ten minutes earlier, Japanese Navy Commander Mitsuo Fuchida relayed his infamous coded signal to the Japanese fleet: TORA! TORA! TORA! The message signified Japan's attacking aircraft had achieved complete surprise over Pearl Harbor. As bombs fell and sailors perished, as torpedoes continued slicing paths through the shallow harbor seawater before slamming into the sides of anchored American warships, Roy glanced up and saw the destruction on Hickam Field. He recalled, "We were on the final approach at less than 600 feet when I spotted at least five aircraft enveloped in flames on the ground. I knew then my country was at war."

As if to emphasize the obvious, two Japanese fighters with guns blazing jumped on the tail of the hapless B-17. Roy said, "We were unarmed and low on fuel. Tracer bullets riddled our plane and ignited the pyrotechnics, then smoke started pouring into the cockpit. Our first instinct was push to full throttle and hopefully escape into cloud cover, but with flames licking the back of our seats, well, Captain Swenson and I realized our only choice was to attempt a landing."

Roy continued, "We could barely see outside because of all the smoke. The bomber bounced so hard it took both of us on the controls to keep the wings level. Then the tail hit. The Fortress buckled and collapsed. When a piece of airfield machinery tried to push the wreckage off the runway, the Fortress broke into two distinct pieces. A photo of the front half of the B-17 is the one people are familiar seeing in history books."

NOTE: Their Flying Fortress is acknowledged as the first B-17, and possibly the first American aircraft, shot down in WWII.

Roy's voiced softened. "Our flight surgeon, Lt. Bill Schick, had suffered a leg wound during the landing but managed to clamber out onto a wing with me and Captain Swenson. We had quite a jump to the ground. As we dashed for cover, a Jap Zero dove to strafe us. Lt. Schick took a round to his head, either a direct hit or as others speculated, a probable ricochet. He later died at the hospital, probably because he refused treatment, thinking other people were more seriously wounded and in need of immediate aid."

NOTE: The pilot of the strafing Zero, Takashi Hirano, paid with his own life for the unnecessary shelling. Flying preposterously low, Hirano's propeller blades gouged the runway. Powerless to maneuver his Zero, Hirano crashed and burned.

Sprinting across the airfield, Roy, Swenson, and their Navigator, Lt. Taylor, saw a sergeant handing out weapons inside a hangar. Roy recalled, "We ran inside and each of us grabbed a pistol and several clips of ammo before heading for the rear door. That old sergeant handing out munitions yelled, 'Hey, you guys have to sign for those weapons!' We shouted back, well, uh…I really don't think you can print what we yelled back."

Roy witnessed the ongoing horror of the Japanese sneak attack at one of the hospitals. "The injured and dying just kept pouring into the hospital," he said. "Guys with missing arms and legs, others hideously burned, several screaming in agony, young men calling for their mothers. The surgeons and nurses did their best to treat wounded and dying men but they faced impossible odds. I'm not usually troubled seeing accidents or even death, but that morning was a bit too much. I sat down on the steps in front of the hospital and had to pull my thoughts together before I could get moving again."

Ernest Leroy "Roy" Reid did indeed get moving again. He did whatever was needed to help his countrymen during and after the attack. Roy eventually flew another B-17 to Australia and became a member of the Kangaroo Squadron, so-named for the first Pacific island hopping B-17s that originated in Pearl. From Australia, Roy flew as copilot on reconnaissance missions and bombing runs. Promoted to 1st Lieutenant, then to Captain, Roy received command of his own Flying Fortress. He flew the B-17 out of Seven Mile Airfield on New Guinea and completed 49 treacherous combat missions throughout the Pacific, including several against the Japanese stronghold at Rabaul. Roy received two Silver Stars and the Distinguished Flying Cross.

With a deep passion for aviation, Roy finagled flight time in a P-38 Lightning, P-51 Mustang, F-4 Hellcat, and P-47 Thunderbolt. Asked how a bomber pilot managed to log time in some of the hottest fighters of the war, Roy said, "The fighter jocks wanted to fly the B-17 and I wanted stick time in their fighters, so we switched cockpits for 'orientation' flights." Asked if regulations may have been disregarded for these 'orientation' flights, Roy replied with a grin, "Nope, we gave each other permission."

After the interview, Colonel Reid offered me a guided tour of his study, or actually his 'man cave.' The walls were a military art gallery of every aircraft Reid piloted during an exemplary 20-year career with the United States Air Force. I spotted a large photo of the celebrated British Spitfire. Intrigued, I asked if a similar 'orientation' flight transaction had been worked out with a British pilot. Colonel Reid replied, "Well, not exactly. I stole the Spitfire. But that's another story."

The day after Roy's B-17 split into two pieces on Hickam Field, the crew boarded the broken Fortress to salvage anything useful. Onboard, Reid discovered four bullets imbedded into the protective steel plate behind his seat. He never objected to the steel plates again.

FOOTNOTE: I had the honor to fly with Colonel Reid aboard the B-17 Flying Fortress featured in the motion picture "The Memphis Belle" by invitation of the Liberty Foundation. My deepest gratitude to these dedicated aviators and ground crews helping to preserve aviation history.

Ernest Leroy "Roy" Reid served in the USAF for 28 years and retired as a full bird Colonel. Colonel Reid reported for his final inspection on September 18, 2015, three months after the passing of his cherished spouse of 74 years, Shirley Root Reid. They rest together forever, side by side, at the national cemetery in Canton, Georgia.

THE LITTLE CHURCH THAT COULD

Only one soldier, Allied Supreme Commander General Dwight D. Eisenhower, controlled the leash restraining the massive invasion taskforce of 5,000 warships jam-packed with a quarter million Allied soldiers. Many vessels had already put to sea. Poised on dozens of airfields all over England, 10,500 aircraft waited eagerly for the long-awaited command, "Go!" Tensions were high as men prepared meet their fate, but more perilous, Allied morale was at risk if another frustrating 'stand down' was issued.

Frustration also consumed General Eisenhower and his HQ

Staff of Admirals, Air Marshals, and Army Generals. Disheartened by a dire weather briefing at 0415 on June 5, 1944, the pessimistic leaders met again at 2130 that evening to hopefully receive a vital favorable weather report from the Scottish meteorologist RAF Group-Captain J.M. Stagg. Tens of thousands of lives were at stake. Emotionlessly, Captain Stagg uttered his now famous forecast, "...rapid and unexpected developments have occurred over the North Atlantic."

Mother Nature had granted a two-day window of opportunity. After brief discussions, General Eisenhower issued the order to commence the Allied Invasion of Normandy, "Okay, we'll go." Operation Overlord was on. In the smoggy predawn light, an English coastguardsman standing guard on the Dorset Cliffs watched in amazement as thousands of warships gradually disappeared over the horizon. Returning home, he solemnly told his wife, "A lot of men are going to die today. We should pray for them."

The Grim Reaper proved the coastguardsman spot-on. A lot of brave men did die on June 6, 1944, as did thousands of French citizens. The persistent, distasteful, and unyielding mission of 'liberation' would continue to pilfer lives and maim civilians and soldiers alike. Bravery was common, most incidents unknown and unreported, yet among the suffering and horrors of D-Day there existed stories of compassion and evidence of the good in man.

On the night of June 5 and into the wee early morning hours of June 6, elements of the 101st Airborne Division parachuted into Drop Zone D behind the exits to Utah Beach. One of their vital objectives was cutting the main Cherbourg/Paris road near the tiny village of Angoville-au-Plain. As the American paratroopers dug in and consolidated their positions, two medics of the 501st Regiment, Robert Wright and Kenneth Moore, set up their aid station in the village's Romanesque 12th century church.

Robert Wright & Kenneth Moore, the two medics

Accompanied by a Lieutenant named Allworth, the two medics searched nearby fields for the wounded and dying. The ancient basilica quickly became more than an aid station as Wright and Moore cared for a growing number of critically wounded men on the threshold of death. After stabilizing the soldiers, Wright and Moore left the church to comb the vicinity for more injured men, including Germans.

Combatants from both sides received medical care with equal compassion. Some died, but as an unrelenting battle continued around the church, Wright and Moore renewed their efforts to rescue and treat injured G.I.s and Germans. Outside, the fight eventually favored the numerically superior Germans, forcing the American paratroopers to retreat from Angoville-au-Plain. Notwithstanding, Wright and Moore refused to evacuate and remained behind to continue treating the wounded and dying.

Angoville au Plain

As the Americans retreated, German infantrymen stormed the church and kicked the doors open. In the silence that followed, the Germans slowly lowered their weapons upon seeing their own men as well as American paratroopers under the medical attention of Wright and Moore. A German officer suddenly appeared inside the church. Observing the kindness tendered by the two American medics, he asked if more of his wounded men could be brought into the church for treatment. Wright and Moore agreed without hesitation or trepidation. Obliged, the German officer even ordered his own doctor to support the American medics. As the Germans soldiers left, they posted a Red Cross flag on the doors to properly identify the historic church as an emergency medical facility.

Vicious fighting continued in and around Angoville-au-Plain for the better part of three days. The village changed hands several times while Wright and Moore continued to administer comfort and aid to the wounded and dying. A mortar round pierced the roof and exploded in the middle aisle, re-wounding many of the injured. Composed and seemingly unconcerned, the American medics never skipped a beat.

Blood-stained pew

Dying men were placed in front of the church to spend their last moments on earth near the altar; it seemed like the proper thing to do. Injured soldiers congested the aisles, blood soaked into the wooden pews, and additional wounded and dying kept coming in. A gravely injured French child was brought into the church; Wright and Moore saved the youngster's life.

In the midst of continuing death and misery, two armed German soldiers who had escaped early capture by the American troops descended from the church tower and surrendered to Wright and Moore. The two medics nor any of the wounded soldiers knew the two Germans had taken refuge in the church tower.

On June 8, a fierce assault by the 506th Regiment of the 101st Airborne Division pushed the Germans out of Angoville-au-Plain for the last time. The French village was firmly in Allied control.

Medics Wright and Moore were credited with treating over 80 men and the one child. The town eventually honored the two brave American medics by installing stained glass windows in the church, illustrating their heroics. The cracked flagstone floor in the epicenter of the church, shattered by the mortar round, can still be seen and the bloodstained pews are still in use. A memorial erected in the town square venerates the two medics and the grateful residents renamed their town square 'Place Toccoa' in honor of the Georgia town where the 101st Airborne Division had trained in the United States. Wright and Moore both received the Silver Star for their humane endeavors.

Later in the war, Robert Wright was wounded during combat in Holland and received a Purple Heart. He became the recipient of two more Purple Hearts during the gallant stand at Bastogne.

Robert Wright passed from this life on Dec 21, 2013. Honoring his request, he is buried in the cemetery next to the Angoville-au-Plain Church.

Kenneth Moore passed a year later, on Dec 7, 2014.

"Not every mad doctor lives in a castle surrounded by villagers with pitchforks. Sometimes they live in the trenches, where there's plenty of spare parts flying all around and a pressing need to get inventive with them."

—From "Meet the Medic"

YOU GOT TO BE KIDDING ME

In an event as devastating as WWII, historians pen the facts and figures and either glorify of demonize the figureheads. Rarely are little known incidents mentioned except as footnotes or an attention-grabbing spinoff. To cite just a few:

Albert Goring, the younger brother of German Luftwaffe Leader Hermann Goering, despised the Nazi party and saved hundreds of Jews and political dissidents during WWII.

On December 6, 1941, the Royal Navy submarine HMS *Perseus* hit an Italian mine near the Greek island of Kefalonia. Most of the crew died instantly, but an intoxicated stoker named John Capes had been drinking rum in a converted torpedo rack and didn't realize the boat had sunk until it hit bottom and started flooding.

Incredibly, he found three other stokers who had been injured in the blast but perfectly capable of helping Cates finish off the rum. With the rum depleted, the four inebriated stokers donned escape gear for an attempt to resurface. Capes was the only one who made it. Exhausted, he swam to a nearby beach, then lost consciousness. He was eventually found and taken to a nearby hospital. Capes recovered from his ordeal but hardly anyone believed his story. The wreckage of the HMS *Perseus* was discovered in 1997. The evidence now existed to confirm Capes' story, 15 years after his death.

The 588th Night Bomber Regiment flew wood and canvas 1920s era OP-2 biplanes to strike targets behind German lines. Some of the Russian pilots flew 18 missions a night without radar, no radios, and explosives held to the wings by flimsy wire. Lt. Col. Papova flew 852 missions and was shot down several times, yet survived the war. The Germans called these infuriating Russian aviators Nachthexen, meaning "The Night Witches." Lt. Colonel Papova and the 588th pilots were all women.

Former Governor of Alabama and 1968 Presidential candidate George Wallace flew on B-29s during WWII as a flight engineer. After the war, Wallace received a monthly VA allowance for his 'nervous disability'. The disability: a fear of flying.

Winston Churchill was famous for his quick wit. At a dinner party, the German Foreign Minister Joachim von Ribbentrop brashly claimed, "The next war will be different, we will have the Italians on our side!" Churchill grinned and said, "That's only fair...we had them last time."

A Yugoslavian man, Dusko Popov, was a double agent in

WWII who hawked his trade under the code name: *Tricycle.* He fed German Intelligence misleading information supplied to him by British Intelligence. Popov also warned the United States of the plan to attack Pearl Harbor, but the warning was ignored. A big gambler, a boozer, and notorious womanizer, Popov was a high-roller living the good life. An Aide to the Chief of British Naval Intelligence applied Popov's lifestyle to the fictional character featured in his spy novels after the war. The Naval Aide was Ian Fleming; his fictional character, James Bond.

I've had the honor to conduct over 300 interviews with military veterans from all branches and all ranks. One story may expose unknown bravery or new information, one veteran's grief or another soldier's joy, yet occasionally a story falls into the 'you-got-to-be-kidding-me' category. This is one of those stories, an incredible account of a B-17 pilot I interviewed in 2012. As with several mindboggling stories from WWII, his is no exception.

On December 7, 1941, Jim Armstrong and his classmates were listening to the radio on the second floor of Harrison Hall at Georgia Tech University in Atlanta. They realized the attack on Pearl Harbor meant the United States was now in a world war and the boys in Harrison Hall would soon be exchanging their books for bombs and bullets. Armstrong attempted to join the Civil Air Patrol. He recalled, "They turned me down because of high blood pressure, so I tried to join the Army Air Corps. Well, I passed the physical with flying colors. Go figure."

As with all WWII pilots, Armstrong trained on numerous planes before selected to command a B-17. He said of the B-17

Flying Fortress, "It was a great bomber, easy to fly, and easy to land. I even buzzed my mother's house just south of Tampa. She was out in the backyard waving at me, very proud of her pilot son. However, I was flying just a bit too low and sort of clipped a few pine trees. It scared the hell out of my instructor, but we made it back okay."

Jim Armstrong

Armstrong went to war with the 8th Air Force, 547th Squadron, 303RD Main Group at Grafton Underwood in Northamptonshire, England. He said, "We flew a few milk runs with P-47 fighter escort but then the morning came for the 'big one', Hamburg, Germany." Over Hamburg, Armstrong's B-17 bomb bay doors jammed and wouldn't open. He said of the incident, "I had no intention of returning to England with a load of incendiaries so I pulled the Red Ball." The Red Ball is positioned on the left side of the pilot's seat. If need be in case of an emergency, the pilot can release the bomb load from the cockpit. Armstrong stated, "The weight of the bombs knocked open the bomb bay doors and we hit the target, but we paid a horrible price for the raid. We were stacked three-high, a low flight, a middle flight, and a high flight. We lost all the B-17s in the low flight."

On being under a fighter attack: "The pilot had to stay focused and trust the gunners to do their job. We flew tight formations but that created a danger of mid-air collisions, which happened too often in my opinion. The German fighters liked the head-on approach, sometimes lining up six abreast then coming straight at us. Our only defense was the top turret gunner. Later in the war, they developed the chin turrets but we sure as heck didn't have them. We even tried to change our tactics, going into the target at very low altitude. It didn't help one bit. You prayed a lot, I remember that."

Armstrong and his crew flew into aviation history on August 17, 1943 as participants in the first raid to the aircraft factories at Regensburg and the huge ball-bearing plant at Schweinfurt. "It was a nightmare," he said. "We lost 60 B-17s and their crews. After our bomb run, a German fighter shot us to pieces, bullet holes everywhere and chunks of metal flying off the Fortress. I had to feather a prop, but we made it home on three engines. We lost a lot of people on that mission, a whole lot."

On September 6, 1943, 338 B-17s bombed Stuttgart, Germany. The raiding force lost at least 45 airplanes, including their crews. None of the bombs hit their target. German fighters ambushed Armstrong's B-17 after his bomb run. He recalled, "They raked us over real good. I tried to reach cloud cover but didn't quite make it."

A fire started behind the pilot's seat, singing Armstrong's hair and burned his face and hands. A waist gunner, Olen Grant, lay in a pool of his own blood after a round pierced his right temple then exited through his right eyeball. The turret gunner tried to harness a parachute around the waist gunner so the body could be tossed out and hopefully recovered so his family could have the gift of closure, but it was not to be. The turret gunner took a round through his head and died instantly as German fighters continued to swarm like an angry hive of killer bees.

With dead and injured still aboard, on fire and trailing smoke, the crippled B-17 was going down in German-occupied France. Armstrong trimmed the heavy bomber for level flight and ordered his crew to 'hit the silk' over enemy territory. Then Armstrong bailed out. In the confusion, two flyboys were left for dead aboard the B-17, the top turret gunner and waist gunner, Olen Grant.

Armstrong said, "We were already dangerously low when we parachuted out so I was able to watch the bomber glide down. She never banked or nosedived and stayed level as if making a normal landing. I couldn't believe it. She made a perfect belly-landing with wheels up in a sugar beet field next to an airfield near the Normandy town of Etrepagny."

The Germans dispatched troops and a French ambulance to the scene, thinking an American pilot had safely crash-landed the B-17. They would search for possible survivors and any potential Intelligence material.

At the crash site, French firemen extinguished the flames before the Germans began their search. Incredibly, the waist gunner, Olen Grant, was still alive. He was helped to his feet then Grant walked out of the B-17 under his own power and straight into the waiting ambulance. He was bleeding profusely. His right eyeball dangled on his right cheek, the pain horrific. German soldiers escorted Grant to a local hospital where he was treated for his injuries. Stabilized, he was moved to a German Air Force (Luftwaffe) hospital in Paris.

Grant lost his right eye but otherwise achieved a full recovery and eventually returned to Allied forces in a prisoner swap. As of this writing, he still resides in Hot Springs, AR.

Armstrong's own journey is worthy of a book. Initially fed and concealed by French villagers, he later walked nonchalantly yet undetected for 50 miles in the direction of Paris. Ironically, one family who gave Armstrong a change of clothes, milk, butter, an apple, and a satchel of food, was Russian.

The French Resistance soon located and protected Armstrong as they continued the journey to the coastal town of Douarnenez. At Douarnenez, Armstrong was joined by 31 other Allied airmen for a boat ride back to England.

Armstrong stated, "I'll always be grateful to the French people. Many of them died helping Allied airmen." Armstrong retired several years ago after a long career as a Presbyterian minister.

Jim Armstrong in 2012

Olen Grant, shot through his temple and right eyeball, survived an unmanned B-17 crash-landing, lying unconscious in a pool of his own blood, taken captive yet humanely treated and cared for by his sworn enemy, was granted his freedom in a prisoner swap and given the God sent opportunity to live out the rest of his life...well, according to Armstrong, "He's still an atheist."

No, I'm not kidding you.

NOT WITHOUT MY BROTHER

Genevieve Sullivan's boyfriend, Bill Ball, lost his life during the Japanese sneak attack on Pearl Harbor. On January 3, 1942, her brothers, George, Frank, Joe, Matt, and Al joined the Navy to avenge his death and serve their country in WWII, with the prerequisite they stay together as a family. Although a policy was in place by the Navy to separate family members from serving together, the rule was never enforced with any authority.

All five brothers were assigned to the USS *Juneau*, a light cruiser that fought in numerous naval battles during the Guadalcanal Campaign. During another engagement on November 13, 1943, the *Juneau* was struck by a Japanese torpedo and withdrew from the battle with other surviving warships. Still seaworthy, she and other vessels headed for the rear-area base of Espiritu Santo for repairs. Below in the murky depths, Japanese submarine I-26 stalked Allied vessels.

It is believed one lucky torpedo from I-26 may have hit the *Juneau's* ammo magazine. The light cruiser exploded and sank rapidly. So fast was the sinking that the senior officer of the task force believed no one could have survived the detonation and ordered the remaining ships to continue on to Espiritu Santo. A B-17 was sent to search for unlikely survivors. In fact, over a hundred sailors had gone into the water.

Due to a SNAFU assortment of mishandled paperwork and reportage, a full eight days passed before a Catalina flying boat picked up ten survivors. The other abandoned sailors had died from exposure, thirst, wounds, starvation, and shark attacks.

All five of the Sullivan brothers were gone. The story told by the men rescued was heartbreaking: Joe, Frank, and Matt were killed in the explosion. Al was said to have drowned a day after the ship sank. George survived for approximately five days until hallucinations, or perhaps grief for his lost brothers, caused him to slip over the side of a raft, never to be seen again.

The brothers' father, Thomas Sullivan, was leaving for work on January 12, 1943 when three men approached the front door. A lieutenant commander delivered the dreadful news, "Sir, I have some news for you about your boys." Thomas Sullivan is reported to have asked, "Which one?" The officer replied, "All five."

Thomas and Alleta Sullivan of Waterllo, Iowa, became our country's only five-star Gold Star family.

Alben and Gunda Borgstrom of Garland, UT, saw five of their sons either join or be drafted during WWII. Two years after the Sullivan disaster, four of the five Borgstrom brothers lost their lives within a five-month period. The Borgstroms were the only four-star Gold Star family during WWII.

Both horrific tragedies prompted the American military to establish new rules and regulations to spare other families such bereavement. The procedure was called the Sole Survivor Policy, as portrayed in the movie *Saving Private Ryan.*

"REUNITE MY BOYS!"

The Morris twins, Bill and Jack, had never been separated until their country entered WWII. Infuriated that her boys had been split up, Mrs. Morris wrote a blistering letter to the President of the United States and the Secretary of War, demanding that her boys be reunited. Mrs. Morris got her way.

Bill learned about Pearl Harbor while attending a movie theater. He recalled, "They flashed the news on the screen about the sneak attack then Uncle Sam suddenly appeared on the same screen pointing his finger at us saying, 'I need you!' Well, when we turned 18, he got us."

The twins were born in Bethlehem, GA in 1924. Bill and Jack, plus three of their brothers and two brothers-in-law served in WWII. Bill said, "Jack and I were inseparable until we joined the Army. They sent me to Camp Crowder, MO and Jack to California. Our mother was furious. She wrote a letter to President Roosevelt and Secretary of War Stimson demanding we be reunited. Mom got her way. Jack and I fought the entire war side by side."

The Morris twins were trained to string communications wire and mastered heavy weapons at Fort Leonard Wood, MO. "It's not Missouri, it's the state of Misery, if you ask me," Bill said.

February of 1944: The twins ship out aboard a Liberty ship for Birmingham, England. Bill said, "We shipped out with a 2½ ton truck, our equipment, and I had a case of mumps. I didn't report to sick bay until we set sail. Jack and I were staying together, no matter what!"

Once in England, the twins underwent specialized training before boarding an LST to cross the English Channel. The date: D-Day plus one, June 7, 1944. Bill recalled, "When we arrived the

LST anchored off Omaha Beach. Shoot, I bet there was at least 5,000 vessels offshore. That night every gun in the fleet opened up. Don't know what they were shooting at, but they sure were shooting at something. Jack and I had been taught to 'get low and get in a hole' but let me tell you something, there ain't no foxholes on a ship! The firepower we witnessed that night was awesome."

Jack & Bill Morris

The next morning, June 8: The twins, their crew, and a 2½ ton communications truck hit Omaha Beach. Bill vividly recalled the scene, "The bodies upset us, hundreds of them, stacked like cords of wood. But we had our job to do so we started stringing wire beyond the rocky bluffs. We hadn't been working long when a group of Army engineers told us to 'back off' to a safer area. Apparently we were working near a mine field. We moved back far enough, I suppose, before things got real ugly. A mine detonated and killed all the engineers."

Night brought no respite. Bill said, "Before dusk I posted guards but the continuous gunfire kept us awake all night. We slept in foxholes with tents pitched over them. I reckon we were a bunch of mavericks, stringing wire wherever we were told to go. Truthfully, I guess we looked like a bunch of mavericks, too, since we had cut our hair into Mohawks for the invasion."

The twins and crew strung wire from Omaha Beach to the Cherbourg Peninsula. They dodged lethal German 88mm artillery and sniper fire, u-turned north through France, and ended up in Belgium. "We slept in ditches, mud holes, hid behind trees, seldom even saw a town, but in Belgium they put us up in a castle that Kaiser Wilhelm used during WWI. That was our first dry floor in three months," Morris said with a grin.

December, 1944: While stringing wire in freezing cold and heavy snow near the German border, Bill and his crew heard the frantic voice of an American officer, "Move out, move out, we're being overrun!" German Panzer tanks were on the prowl, the Battle of the Bulge was in full swing.

Bill said, "We hustled back, but recruits fresh out of basic were ordered to remain in place and try to stop the Panzers with ineffective bazookas. Those poor guys didn't make it out."

Dense overcast skies, below-zero temperatures, and heavy snow worked to the Germans' advantage. Bill stated, "Our planes were grounded and that lack of air cover cost a lot of American lives. Then, on December 23, we awoke to a clear blue sky. Here they come, thousands of them, filling up the clear blue sky with their contrails. Our flyboys got the job done; and by the end of January 1945, the Battle of the Bulge was over."

Bill remembered a peculiar incident while stringing wire on the German side of the Rhine River. "We were on the river bank when suddenly bullets started 'pinging' off the brick wall behind us. Well, just like we'd been trained to do, we got low and got in a hole!" American sharpshooters on the opposite bank were shooting at German mines floating down the Rhine River. Bill said, "Those boys were either lousy shots or their bullets were ricocheting off the water. Let's hope it was ricochets!"

The twins and crew were with American forces when they liberated the infamous concentration camp at Buchenwald. Morris softly recalled, "We could smell the camp from a mile away. My God, what looked to be humans stumbled around like scrawny zombies. It was hard for us to believe those poor souls were even human, much less alive. Then we found the meat hook conveyor line that ran bodies to the furnaces and we found trenches filled with… well, I guess that's enough on that subject. I've tried to forget that place, Buchenwald, but it's impossible, and the people who deny the Holocaust are foolish. It happened, I know, I was there."

After Germany surrendered, Bill and Jack and their truck crew boarded a troop ship in Marseilles then set sail for the South Pacific island of Okinawa. Bill explained, "We were there to prepare for the invasion of Japan, and I might add we weren't too thrilled about it either. That first night we had to sleep on the beach. We didn't think sleeping on the beach was a big deal until the next morning when 75 Jap soldiers walked out of the jungle and surrendered to us. They carried American cigarettes, C-rations, and a bunch of American hand grenades. I'll tell you one thing, we sure were glad they were in a mood to surrender!"

The lives of Bill and Jack Morris, their crew, feasibly over a million other American casualties, and an estimated 10 to 20 million Japanese deaths were spared by the two atomic bombs. Bill said, "Both wars were finally over. Jack and I were ready to see Georgia again."

After arriving in Seattle via troop ship, Bill and Jack took a train across America. From Chattanooga, all the way into Atlanta, the locomotive blew its whistle full blast in celebration. Bill said, "We hitch-hiked a ride to Monroe Street in Social Circle then walked home. I still remember our mother's welcoming words to this day, 'I'm so proud of you boys.' Well, I guess we did what we had to do, and it was over."

His closing comment: "I hope and pray that America never forgets the sacrifices made in WWII. We saw stacks after stacks of bodies at Hurtgen Forest, and I'll never be able to forget seeing the paratrooper who crawled out of a flooded field behind Omaha Beach where he died in place. And a man's jet black hair turn white in just two days. As they say, war is hell, and it's the truth. I saw too much and have too many memories, but the Lord let me live, and I live for Him."

All five of the Morris boys and two brothers-in-law made it home safely. In 1998, 79 surviving veterans of Company B 32nd Signal Construction Battalion attended their reunion. In 2008, only seven were in attendance. Jack Morris reported for his final inspection in 1993.

Bill followed Jack on November 11, 2015 to report for his final inspection. Rumor has it, the twins are stringing wire from the Pearly Gates to beyond.

*****

The following poem was penned by the daughters of the veterans from the 32nd Signal Construction Battalion for one of their reunions:

It was 50 some years ago
In this same location
That a group of young men came
From all over the nation
These lines are a tribute
To the impact they had

In the war that they fought
And to the one that's our dad
It started with training
For company B and for A
To learn to string wire
In all kinds of ways
They learned to work in the dark
And how to dig holes
To forget that you're tired
And how to climb poles
Then they sailed out the harbor
Where some sober young men
Didn't know if they'd see
The Statue of Liberty again
They landed in England
For training some more
And were on Omaha Beach
By June of 1944
The war was intense
And the going was slow
'Til they finally broke through
At a place called St. Lo
Then on through France
Where it seemed like a race
To keep pace with Patton
And his incredible pace
All across Europe

They strung spiral four
They fixed it and spliced it
And then strung up some more
They hung it on bushes
And on buildings too
All to make sure
That the message got through
The Battle of the Bulge
Gave reason to dream
Of a white Christmas
And what it would mean

To be back home
With family and friends
And a life that is normal
And a war that would end
But for now they must settle
For sending V-mail
It took a long time
It was nothing like E-mail
The bridge at Ramagen
The first place on the Rhine
That the river was crossed
With a lead cable line
When Hitler surrendered
And gave up on May 8
The 1st Army had thought

They could now celebrate
But soon they found out
The end wasn't in sight
The war with Japan
Had more battles to fight
So off through the canal
Sailed 5000 men
Not knowing what would
Be happening to them

Across hot miles of ocean
On a miserable trip
The sick ones just wanted
To get off the ship
V-J day had happened
You'd think now for sure
It'd be time to go home
And forget about war
The Okinawan typhoons
Gave them all quite a scare
The noise of the mongoose
Made them say "who goes there?"
They finally came home
And started their lives
In peace time surroundings
With children and wives
Which brings us the subject

Again of our dad
We sure are thankful
That he's the one that we had
He took us on trips
And we camped in a tent
We had lots of fun
All the places we went

He wrote us all poems
When we were just small
And it looks like we turned out
Like him after all
He coached our ball games
Taught us to catch, throw and hit
That we were all girls
He didn't mind a bit
He paid for our weddings
And the cars that we crashed
For a much better dad
We couldn't have asked
But far more important
Than this temporal stuff
In his godly example
Of a man we could trust
To be fair and honest
And mean what he said
And to not compromise

But do the right thing instead
We all can remember
Our dad and our mom
On their knees by their bed
Praying for what we'd become

So, now in conclusion
We want to honor you men
And the ones who aren't with us
And we won't see again
Tom Brokaw was right
When he gave you a name
"The Greatest Generation"
Is what you became
So tonight, to you veterans
And especially our dad
We bless and salute you
For the courage you had.

BRAVERY BENEATH THE WAVES

My father fought in the CBI (China, Burma, India) theater of operations with the Army Air Corps during WWII. As far as I know, dad returned home physically and mentally intact. My Uncle Melvin joined the U.S. Marines and survived some of the most brutal battles in the Pacific.

Our clan in Memphis claimed Uncle Melvin was never the same after the war. He died of a self-inflicted gunshot wound later in life. My Uncle Thomas found himself in the middle of the Battle of the Bulge, survived ruthless hand-to-hand combat and recovered from frozen feet blacker than two lumps of coal. He eventually started his own construction business and built some of the first modern casinos and hotels in Las Vegas.

Most of the men in my father's clan served in WWII, as did my mother's. It was the 'right' thing to do. From hearing their stories around the dinner table or listening to several late night B.S. sessions, my interest in military history and WWII sprouted from an early age. I read everything I could get my hands on or shoplift (only once, got caught, and had my rump blistered by loving parents who were still allowed to discipline an unruly son). I watched and learned and memorized war movies and absorbed the information from documentaries like *The Valiant Years*, or weekly TV programs like *Combat* or *Twelve O'clock High*. But my favorite series was *Victory at Sea*.

Here's the thing: *Victory at Sea* was an outstanding tribute to the naval history of WWII. The series received abundant accolades, including award-winning music and soundtrack by Richard Rogers. But for me, as much as I admired and enjoyed *Victory at Sea*, I was never inspired to navigate around the world as potential shark bait. Terrain is for humans, oceans for hungry fish, many with attitude. Even huge WWII aircraft carriers fought for survival against an angered Mother Nature in the configuration of a typhoon or hurricane. However, typhoons and hurricanes are usually predictable; rogue waves are not. A wall of seawater 50 feet high with the power to capsize a military vessel has my deepest respect…from a distant dock.

An extraordinary breed of Americans fills the ranks and vessels of the United States Navy. I just ain't one of them. With all due respect to the courage and devotion of sailors from all navies, there's not enough Dramamine on the market to make me sail the Seven Seas, especially in time of war. It's hard to believe, but a few selected anthropoids even delight in serving UNDER the sea. They are known as Submariners.

The Collegiate Dictionary describes a submarine as: a ship that can submerge and navigate underwater, or, something situated and living under the surface of the ocean, as an animal or plant.

"…something situated and living under the surface of the ocean, as an animal or plant." I believe those words simply tell us that we don't naturally belong there. Submerged can also mean sunk, and the word 'navigate', well, dolphins and whales are naturally equipped to navigate underwater but humans are not! There are no fire escapes on a sub, no visible lifeboats, no parachutes, and no place to go if things quit working, except DOWN! WWII submarine toilets even had a thingamajig apparatus that if improperly set would discharge the toilet contents right back at you instead of flushing the contents DOWN, which brings up the thought-provoking question: Why the heck are Dress Whites Navy attire?

Charles Crews exemplified that extraordinary breed of man willing to fight a war beneath the waves. A battleship was too big with too many people for Charles; he liked the team concept, small and personal, he preferred the submarine. I interviewed Charles Crews in his home, a ranch-style brick dwelling, neatly kept and neatly trimmed—just like the owner.

CHARLES CREWS

Charles Crews fought WWII beneath the Pacific Ocean as a submariner aboard the USS *Spot*. The *Spot* tallied a notable war record of sunk and destroyed enemy targets, including one radar installation. She was depth-charged four times, twice by her own country.

Born in 1921, Crews said, "I lost both parents before my 14th birthday. We kids were raised by a compassionate stepmother, but I was on my own at 18." Crews milked cows and delivered milk until introduced to his civilian niche as a projectionist in movie theaters. "A friend that operated projectors in the Navy taught me the trade," he said.

Captivated by his friend's inspiring tales of the Seven Seas, Crews tried to join the Navy in 1939. "I was turned down," he said. "Seems like I had flat feet." Crews found work at the fabulous Fox Theater in Atlanta. "I worked as an usher before receiving a promotion to projectionist. Part of my job in the screening room was censoring films, like trying to cut 'damn' from *Gone with the Wind*."

The Navy overlooked Crews' flat feet after the Japanese ravaged the 7th Fleet at Pearl Harbor. Crews, an Atlanta native, recalled, "I boarded a train at Union Station in Oct '42 for boot camp at Great Lakes, IL. Our barracks was built in a corn field, no hot water, no heat, but a whole bunch of Yankees. We refought the Civil War every day."

Crews' first shipboard assignment was on the refurbished survivor of December 7, 1941, the battleship USS *Nevada*. "I boarded in Seattle," he said. "We took trial runs on Puget Sound before sailing to Long Beach, CA. I danced with Betty Grable at a Hollywood eatery, saw Clark Gable and Bob Hope, and heard aging Sophie Tucker tell the boys, "There's snow on the roof but there's still a fire in the furnace."

The *Nevada* joined battleships *Idaho* and *Texas* to bombard Attu Island in the Aleutians in '43. As a 40mm gunner, Crews watched the massive 16" guns fire their payloads. "We put cotton in our ears," he said. "The battleship would rock like a baby cradle when those guns cut loose." From Attu, the *Nevada* headed south along the West Coast then slipped through the Panama Canal. Her next port-of-call was the lush green countryside of Ireland. Albeit, Crews soon realized that battleships were not his forte. He said, "I liked the idea of a closer-knit unit." Crews volunteered for submarine duty and passed the written and physical tests, including the prerequisite of having all his teeth.

He reported to New London, CT for submarine training and survived the 100ft diving tank, learned to breathe underwater with the Musson Lung, endured a pressure tank (a 50% failure rate), mastered 'the boat' from bow to stern, and trained at sea on antiquated WWI subs. Graduating 6th in a class of 150, Crews received orders for the USS *Seawolf*. Crews said, "The parents of a boy from Massachusetts wanted him to remain on the east coast so I swapped boats with him. I took the USS *Spot*." (The USS *Seawolf* was lost at sea during Oct of 1944. There were no survivors).

Charles Crews, U.S. Navy

Crews joined the crew of the recently commissioned USS *Spot* in San Francisco. He said, "My job was operating the starboard side maneuvering board and keeping fresh water in the batteries. A WWII sub used diesel engines when running on the surface but used batteries if submerged. To water the batteries, I had to crawl down a hole with a hose. Leaking acid was always a big danger." Ordered to Pearl Harbor, the *Spot's* crewmembers received a big surprise upon their arrival. Crews recalled, "We were told that we'd been sunk. That sure was news to us!"

Submarine USS Spot

Combat lay ahead. Crews recalled, "Wake Island was our first patrol. We sank merchant ships but I don't remember any celebrations. There weren't any feelings of guilt, but we weren't jovial about it either." The USS *Spot* made occasional unauthorized rendezvous with Japanese 'Junks'. "The Junks were everywhere," Crews said.

"We knew they could radio in our position but we still traded with them, usually our canned goods for their fresh fish. The Navy finally stopped the risky barter system."

Armed with torpedoes, a 5" deck gun, 40mm and 20mm anti-aircraft guns, Crews spoke of one encounter between the *Spot* and an enemy merchant ship. "Our 5" gun traded shells with the enemy ship. We won the fight then sent a boarding party to search the vessel. What the guys saw on the enemy ship made them sick; the carnage was horrible, they wouldn't even talk about it."

While taking on supplies from a sub tender off the island of Saipan, the sailors on *Spot* witnessed the grisly results of fanatical militarism. "Bleached bones of suicidal Japanese dotted the beach," Crews said. "We saw other bones, too, American."

The *Spot* received credit for participation in the battle for Iwo Jima. Crews said, "We remained on station to rescue downed airmen, young flyboys like future President George H. W. Bush, but other subs always beat us to the rescue." (Bush was shot down and rescued by the submarine USS *Finback* near the island of Chichi Jima).

The Yellow Sea near China and the waters off Guam were also patrolling territory for the *Spot*. Crews recalled, "On one occasion we surfaced inside an enemy convoy and used our 5" deck gun to sink a couple of merchant ships, then submerged. We thought the fighting was done, but about the time we were settling into bunks the order barked over the boat's intercom, 'Man your battle stations!' The skipper was resurfacing to engage another merchant vessel. Well, he had made a huge mistake. It wasn't a merchant ship, it was a well-armed Japanese mine sweeper. It engaged our gunners on deck and raked them over pretty good. Several guys were badly wounded. We submerged immediately, right into a mud bank. Depth charges pounded us; it's like bombs going off in your face. I prayed for God to help us, and He did."

Below deck, the pharmacist mate tended to the wounded. Crews recalled, "I was trying to help Doc but when he told me to stick a guy with morphine I didn't know what to do. Doc yelled at me, 'Damn it, Crews, stick the man!' Well, I stuck him, and did okay, I guess." Due to mistaken identity, the *Spot* was attacked by the Army Air Force and once by a US Navy destroyer. "I'll tell you this, that got old quick," Crews said. Crewmembers of the US destroyer apologized to the *Spot's* crew at a recent reunion. Crews stated, "To be honest, I didn't want to hear it!"

On the popularity of sub grub, Crews stated, "Well, we had steaks, plenty of steaks, but we longed for anything green, like spinach and lettuce. When in port we 'pilfered' crates that had anything leafy green inside."

In early 1945, the *Spot* found herself in trouble in the Sea of Japan. Crews recalled, "We surfaced right in the middle of a mine field. The skipper screamed, 'Shut down engines, shut down engines!' I can't describe my feelings, especially when he said, 'Follow orders and be on your toes.' I prayed that day, too, pretty much all day. Then we heard 'Dive!' and submerged straight down, like a brick, and we luckily slipped away."

In need of provisions and refurbishing, the *Spot* returned to Pearl Harbor in July, 1945. One month later, two atomic bombs dropped on Japan made the expensive overhaul and resupply unnecessary. As Japan surrendered and another World War ended, Charles Crews was leisurely sunbathing on Waikiki Beach. He implied with a grin, "That's a great way to end a war, don't you think?"

Crews worked as a film inspector for the Civil Service until his retirement in 1997, then worked independently for and with audiovisual support systems. "My last gig, in 2009, was a Mercedes Benz trade show at the World Congress Center in Atlanta," he said. "I believe I was paid five grand for the one week. My first paycheck as a projectionist was $12.00. Change can be good!"

ALL IN THE FAMILY

My father's condensed story first appeared in newspapers way down south in the lush Peach State of Georgia. That said and done, while working on the longer version for aveteransstory.us, at first I considered researching additional battles and incidents in the China-Burma-India Theater, my father's port-of-call in WWII. A foray into our squirrel-infested attic changed my mind.

Rummaging through our attic ought to qualify an individual for combat pay, but I have yet to receive a response to my request from the Department of Defense, much less a check. Our attic looks like a setting for '*The Walking Dead*' but I'm guessing after one glimpse at the mess the zombies would drop dead, pardon the pun. I'll spare readers further details, but I hope the producers of the TV series *'Hoarding, Buried Alive'* never receive a clandestine tip on the squirreled away, pardon the second pun, junk at the top of our staircase.

Anyway, while digging through a dusty color-coded tote, I found my father's army uniform, campaign ribbons, decorations, an ivory letter opener, sergeant stripes, additional military records, coins of unknown value, and the coolest thing of all, a scrapbook filled with photos of dad's military days I'd never seen before.

For some reason, I grew up thinking my father served less than two years in the CBI Theater of Operations, but I discovered a yellowed newspaper clipping that claimed dad served over 30 months. Another tote contained information on his Asiatic-Pacific Theater Ribbons, which included Burma, India-Burma, China Defense, Central Burma, the China Offense Campaigns, and many more. Those decorations made sense.

What I didn't expect was to discover the EAME Ribbon, which covered European, African, and Middle Eastern Campaigns. I knew my father hopped, skipped, and jumped his way to Burma aboard a C-47 Gooney Bird and I also remember he mentioned something about a 'lengthy stopover' somewhere in Africa. Thing is, I never heard the details, or perhaps I was too young then to recall now.

And most proudly, in the same newspaper clipping I found out for the first time that my father was awarded, not one, but **two** Bronze Stars. Dad never mentioned the Bronze Stars but I'm fairly certain I know why. He was not the type of man to brag. Rather, he tried to find humor and goodness in his fellow man and the day-to-day world we refer to as the Rat Race.

Sgt. Peter Jay Mecca, Sr.

Dad always spoke of China or Burma or India in a positive way, with the exception of the dire poverty. He hated the lack of even basic living conditions for the populaces while a "select few" lived in opulence. My father was not an activist-type person who believed he could change the world; he was a realist, but he had the gentlest heart of any man I've ever known. He saw the good in people, not their wicked ways, not their religion, not their color… he judged their character.

Dad loved his country. He was a patriot, a pragmatist, a good son, brother, uncle, husband, soldier, and one hell of a great father.

ITALIAN, AND PROUD OF IT

Peter's father, Leonardo, was born on March 13, 1863. Leonardo's future wife, Anna Maria, was born on June 29, 1868. Both of Peter's parents came from a small poverty-stricken mountain top village in Italy called Avigliano. Leonardo sought a better life for his wife and the 11 children he and Anna Maria would procreate. That better life beckoned from the United States of America. Leonardo came over on a boat and saved his money until he could afford passage for his wife and their growing family. The baby of the family, Peter, was born on Feb 27, 1907 in a coal mining community on the outskirts of Scranton, PA, a little town called Dunmore.

Bilingual, Peter spoke fluent Italian and English, but Leonardo constantly told his youngest offspring, "You're an American, so you speak only English." Leonardo demanded the same of all his children; "speak only English", although Leonardo never learned to speak the new language.

Peter grew up in this new world surrounded by a large family of nephews and nieces and cousins, aunt and uncles from his mother's clan, the Varrastros, and brother-in-law Dan's clan, the Coluccis. Patriotic and hard-working, the Italians in the Scranton/Dunmore vicinity were not wealthy, but being Italian, there was always food on the table, especially a big bowl of spaghetti. Anyone was welcomed to a plate of all-you-can-eat pasta or perhaps offered a snack of pan-fried Rainbow Trout caught from an ice-cold Pocono Mountain stream.

Their houses were heated by coal; deep snow was no excuse to miss school. Breakneck speeds on a sled zooming down "The Hill" in Dunmore broke many a young bone, but it sure was fun. The local baseball field became *A Field of Dreams* for many a youngster wanting to follow in the cleat-prints of beloved New York Yankees like Ruth and DiMaggio, Berra and Mantle, a few local ball players actually fulfilled the dream.

Peter was a ball player, a trout fisherman, a sharpshooter of a hunter, and in time made a living as an electrician. Like many Italian men in pre-World War Two days, Peter was still living at home with his extended family when the bulletins on Pearl Harbor hit the airwaves. America was at war. On Jan 19, 1942, Peter enlisted to serve his country in the Army Air Corps.

Designated for O.C.S. (Officer Candidate School), Peter's officer training was terminated after the Army found out his civilian occupation was that of electrician. Told to sew on sergeant's stripes, Peter received inoculations for small pox, yellow fever, tetanus, cholera, typhus, typhoid, and one that told volumes of his next port-of-call: Indian Typhoid. Peter was headed for CBI, the China-Burma-India theatre of operations.

Transferred to the 100th Transportation Squadron, 1st Ferrying Group, the precise location of his base is debatable, conceivably Mohanbari or Sookerating, but most likely Chabua. All three of the airbases were cut from British tea plantations, and all three bases supported the pilots and aircraft flying the treacherous Himalayan Mountain Range, better known as 'The Hump.'

Army records indicate Peter 'strung wire' throughout the CBI region. He installed telephone lines and equipment, wired buildings and power systems, repaired switch boxes, outlets, and pull boxes, plus supervised nine other men in requisitioning, better described as 'legalized pilfering'. Peter also dodged strafing Japanese fighters and bombers. During a bomb raid he received an injury when another soldier jumped into the same foxhole and landed on my father's neck. Neck pain plagued Peter for the rest of his life. He did, however, obtain compensation from the government: 60% disability and a check for $69.00 a month.

He often told the story of Photo Joe (Foto Jo). The narrative went something like this: "Our reconnaissance plane (most likely a P-38 Lightning) would play tag with the Jap recon plane, pretending they were both armed with something other than a camera. One day the Jap got a bit too aggressive and our recon pilot was angry as hell. He landed, told the ground crew to rip out the camera and install the .50 calibers, and was waiting on Photo Joe the next morning. Well, they played tag again above our base but this time it turned out to be a one-sided dogfight. The Jap went down in flames. We saw the whole thing."

Peter authorized monthly Class B Allotments of $18.75 for

the purchase of War Savings Bonds, Series E, to be sent to his sister, Mrs. Grace Colucci. He wrote of his dismay upon witnessing the horrible poverty and living conditions of selected peoples in the underclass of India. Occasionally Peter flew the 'Hump' on C-47 Gooney Birds and C-46 Commandos, but seldom discussed the flights with family or friends other than mentioning how the mountain range was littered with the debris of hundreds upon hundreds of Allied aircraft. Japanese fighter aircraft routinely intercepted the vulnerable cargo planes, calling their tactic, 'Tsuji-gin', meaning 'Cutting down a casually-met stranger.' Army records indicate Peter qualified with the M-1 Carbine as an 'expert marksman' on several occasions, he never missed the target. His son would repeat 'expert marksmanship' using the same weapon with the same perfect score in basic training before his first tour in Vietnam.

Other details of my father's war experiences are too sketchy or memorized hearsay from family members to accurately portray an honest narrative. It is known that after three years of war, my father was sent to Camp Luna, NM before his last military assignment at the Memphis Army Depot on Jackson Avenue in Memphis, TN.

There, in Memphis, the Italian from Dunmore, PA fell in love with a Southern Belle named Lucille and married the beautiful lady after a six-week courtship. Peter was honorably discharged on Oct 23, 1945. He remained in Memphis with his lovely bride and earned a lifelong paycheck as an electrical supervisor for a hardwood flooring conglomerate. He toiled six and seven days a week to provide middle-class housing and a good education for his only son.

Staff Sergeant Peter Jay Mecca, Sr. passed from this life on Dec 27, 1981. I still miss him. This story is for you, Dad.

ROASTED WIENIES, BANANA SPLITS, SOUTHERN FRIED CHICKEN, OKRA, AND ONE BIG RED RUBBER

As our G.I.s swarmed into Jolly Old England during World War II, the British media grumbled, "The fault with Yanks is that they are over-sexed, over-paid, and they are over here!" Conversely, thousands of British men trained in the U.S. during the war. Yet, unlike our boys in England, British pilots were welcomed in the former 'colonies' and were grossly under-paid. Plus, British pilots were not criticized (not often) for chasing the American fairer sex.

After the downfall of France in WWII, the German Luftwaffe controlled the skies over Europe. Their next objective was to sweep the Royal Air Force from unfriendly skies as an essential part of Hitler's pre-invasion scenario. The British government and military were painfully aware their precious resource called 'pilots' was in limited supply and immediately expanded flight training. A big problem arose: with a limited landmass, Britain's airfields were vulnerable to marauding German aircraft.

British aircraft, the graceful Spitfire and reliable Hurricane, were more than a match for the German Messerschmitt Bf-109s and Focke-Wulf Bf-190s. Although the British aircraft were in short supply, they retained a unique edge by utilizing the new-fangled development called radar. What the British could ill-afford was the loss of human resources. The concept arose to train British pilots in 'safe' or 'neutral' countries, including the 'nonaligned' United States of America.

British pilots in training on American soil | Denis Payne, second from right with flight cap

In America, the Arnold Scheme kicked into high gear. Named for its creator, General "Hap" Arnold, the USAAC (U.S. Army Air Corps) instructed British aviators at bases positioned in the SEACTC (Southeast Air Corps Training Centre). The isolated airfields or nearby cities chosen included Lakeland and Arcadia, FL, Camden, SC, Tuscaloosa, AL, Americus, GA, and Albany, GA. At the 'secret' rough but ready airstrips, over 18,000 British pilots earned their wings over American turf. Denis Payne was one of those aviators.

BULLETS AND BUTTER BEANS

Born into a military family on Sept 11, 1919 near Devonport, England, Denis Payne graduated at age 15 from the Colchester High School for Young Gentlemen. After Colchester, young Gentleman Payne attended a Royal Air Force school to study engineering. By the age of 19, he was flying as a flight engineer with an RAF Squadron aboard obsolete Fairey Battles, a single-engine bomber. Venerable and slow, the Fairey Battle was easy prey for German fighters over France at the outbreak of WWII.

Payne stated, "I realized I'd have a better chance to survive the war by crawling out of the backseat and getting into a pilot's seat." He applied and was quickly accepted for fighter training, but before his orders came through, Payne fought in the historic Battle of Britain aboard a Bristol Beaufighter night fighter. He said, "The Beaufighter was a tough little aircraft. I remember those missions, all the bullet holes in our airplane, a bit scary, it was."

In the late summer of 1941, Payne finally received his orders for fighter pilot training…in America. Sent to Montgomery, AL, he recalled, "We spent the first two weeks trying to adjust to the weather and coming to grips with the Southern language. If you recall, Winston Churchill stated, 'England and America are two countries divided by a common language.' All I can say is, the chap knew what he was talking about."

The British boys suffered in the summer heat of Montgomery, but got even hotter under their collars trying to understand the tutoring and instruction of an apparently vulgar system known as West Point military inflexibility. Payne declared, "Bloody hell. We got up at 0500 and hustled everywhere. They told us when to eat; they yelled at us; we were not very happy blokes!" Payne's next port-of-call for advanced training was the 'Secret' Southern Airfield in Americus, Georgia, renamed Jimmy Carter Regional Airport in 2009. Now Payne really had to adapt to the 'Southern' way of life, plus learn a new problematic dialect.

In a futile attempt to maintain 'secrecy', British pilots were not allowed to wear their uniforms into the city of Americus because of American 'neutrality' at the time. Payne stated, "Every man, woman, and child in Americus knew we were British subjects, but the people treated us magnificently. The town invited us to a 'weenie roast' during our first week. We had never eaten hot dogs and had a dissimilar meaning for the word 'weenie', so I guess it's needless to say we were quite concerned about what the town blokes wanted to roast."

While mastering the legendary bi-wing PT-17 Stearman, Payne fell in love with America and its people. "We were born into a rigid class system in England, but in Americus we were treated as equals and with respect. The townspeople were easy-going and approachable, a huge difference from the system I grew up in."

On Southern cooking: "I loved Southern cooking then and I

still love it now. We ate fried chicken for the first time and could get peaches, corn on the cob, black-eyed peas, butter beans, and cornbread. I visited the local drug store and sat down at what they called a lunch counter. I ordered this thing called a banana split. When that treat was placed in front of me I thought I'd died and gone to heaven." Albeit, when asked about fried okra, Payne stated, "Listen, mate, I'm only 94 years old, give me a bit more time."

And his opinion of Southern iced tea? "Now, there's a bit of southern culture that's hard to understand. It's bad enough that southerners drop ice cubes in their tea glasses, but then they add sugar to sweeten it. After that they add lemons to make it tart again; it doesn't make any bloody sense." When asked about American beer, Payne said, "Well, put it this way, we walked into a store in Americus to hopefully purchase a few beers. We were handed bottles of Root Beer. Not exactly what we had in mind. We figured the Yanks were hard up for alcohol."

Payne recalled an embarrassing incident: "I was in town with my mates one day when we ran into the mayor's wife. A nice lady, she was, always had time to talk with British pilots. When I asked what she was doing in town, she said, 'Oh, just piddling around.' Bloody hell, we were speechless. If a British pilot is 'piddling' around it's because he's had one too many at the local pub. I was really concerned about the mayor's wife after that conversation."

A pretty young girl from Manchester, TN, named Mary, found employment in Macon during the war. She and Payne met one weekend while she was visiting in Americus. Sparks flew and Cupid proved to be an expert archer. The couple tied the knot six weeks later and they've been fighting the language barrier for 74 years. Payne said, "When we first started dating, I told Mary to meet me in front of Gaham's the next weekend. Mary said she had no idea what I was talking about. I told her it was the big building in town with the name plainly painted on the front in sizeable letters. Well, the building was advertising Georgia hams, the lettering was GA. HAMS."

Mary Payne automatically became a British subject by marriage. Once in England, she was scheduled for employment with the British military, but an American general scooped her up due to her dictation skills. Mary said with a grin, "I believe my talent on a piano helped, too." Mary had entertained American troops with her musical aptitude while en route to England via a 60-ship convoy.

Breaking bread at the dinner table for the first time after meeting Mary's parents, Payne asked her father, "With your permission, sir, I'd like to knock up your daughter in the morning.' Mary recalled, "Good Lord, dad almost stroked out. His face turned fire engine red." When a Brit asks for the permission to, 'knock someone up in the morning', it's their way of requesting, "May I have the honor to knock on your bedroom door in the morning and serve you breakfast in bed.' For the British, the request is a thoughtful and polite gesture. But for an American, Payne said, "Bloody hell, the fathers could be quite unsettled by the offer."

Payne's piloting skills earned him a job teaching navigation in Macon, GA. One of his responsibilities required correcting map and plotting mistakes made by new recruits. He used a large red eraser to correct the penciled-in errors.

The embarrassing incident: "Well, I corrected so many mistakes I needed a new rubber, that's what Brits call an eraser. I went downtown to the Woolworth's Store and informed an attractive female employee that I needed a big red rubber. She stuttered incoherently for a moment then directed me across the street to the Rexall Drugstore. I didn't understand why, but being polite I just did as she requested.

I approached a female worker in the drugstore and dutifully asked her for a big red rubber. Well, her face flushed, too. I wondered what the bloody hell I was doing wrong. Then she excused herself and fetched the manager. He was a nice lad, but I had to repeat my request. He said, 'Yes, sir, I'm glad to help. Would you like a dozen?' I said, 'Good Lord, man, I only need one!' He said, 'yes, sir', and walked behind the counter. He opened a drawer, reached in, and pulled out a handful of Trojans, in all sizes.

I realized my blooper, but was too embarrassed to say a word, so I just bought one, a big one. Then I walked immediately back to the British barracks, opened the package and took the thing out and pinned it to the bulletin board with a large note: *Warning. This is what the bloody Yanks call a rubber*!"

Payne's abilities earned him an instructor's slot, receiving the same patriotic pitch from his superiors three times during WWII, 'Good instructors are more important than pilots.' Ports-of-calls included instructing cadets in Macon, GA and the twin-engine fighter school in Scotland. Finally assigned to British Bomber

Command, he trained British pilots on safe landing techniques in inclement weather for the Lancaster, Ventura, and Wellington bombers. A natural pilot, his knowledge earned him status as an Aviator-of-all-trades.

June 6, 1944: The Allies launch D-Day, the Normandy Invasion to liberate Europe. Payne said of the morning, "We were taking off in a Wellington bomber and, per procedure, pulled up the undercarriage. We never identified the cause, but the bomber continued on down the runway. She pancaked, skidded down the runway and across a motorway then smacked nose-first into a haystack. A brand new Wellington bomber she was, classified radar and other sensitive equipment, totally destroyed."

That afternoon, Payne and other aviators were celebrating D-Day in the officer's mess. "Well, we had a pint too much," he stated. "We passed around a new but sterile pristine chamber pot filled to the brim with ale. I may say our antics compelled several blokes to leave the bar. Later we staggered outside, physically picked up an MG sports car, and placed it inside the bar. I can't tell you what happened after that because we left before the car's owner showed up."

After VE-Day, Denis and Mary paid a visit to her hometown in Manchester, TN. Payne stated, "Mary hadn't been home in three years and was anxious to see her family. I was very happy for her, but I never thought I'd be arrested for running corn liquor!" Mary's friends packed the trunk of Denis' car with corn liquor then called their high school buddies at the local police station to advise them of the prank. Mary sent Denis on an errand but the law lay in wait.

They pulled Denis over, checked the trunk of his car, of course found the corn liquor, and arrested Denis on the spot. He said, "It was pretty funny after it was over. I still have the distinction as the only British subject thrown into jail for running corn liquor in Manchester, TN."

Denis and Mary Payne's charming sense of humor conceal their graphic experiences of WWII. They witnessed death and destruction on a daily basis, lost several friends, and knew that the grim reaper could make the fatal call at any given moment. Nevertheless, they fought and lived and loved through the inferno of a World War. The war is long over, but the Payne's are still living, still loving, and still laughing.

Payne in front seat of a Stearman biplane, pilot John Laughter in rear seat

Payne's father saw combat in the China-Burma-India Theater; his oldest brother saw action in Africa, his 2nd brother in Egypt, the youngest brother also served with the British army. Payne eventually worked for and retired from the British Consulate in Atlanta, GA. "That's right, mate," he said proudly. "I am the original James Bond, 007. And if you care to believe that, I'll knock you up with a cup of tea."

Gentleman Denis Payne passed gently into the good night on June 11, 2013.

IN THE MIDDLE OF NOWHERE

Best-selling paperbacks, novels, and especially Tinseltown, have a habit of portraying our national heroes or well-known battles of World War II with little concern for lesser known campaigns and casualties. They, too, were heroes; these, too, were hard-fought battles for real estate in the middle of nowhere, and we should honor their suffering and patriotism.

In the Pacific, thousands of young men of various nationalities were shot down, sunk, missing, forgotten, and in the case of countless Japanese, abandoned on lush jungle islands to wither and die. Modern-day tourists and military historians can stroll beaches where men died or may actually walk by an area of undergrowth containing the remains of a fallen warrior.

Pearl Harbor, Midway, Iwo Jima, and Guadalcanal are names well-known to most Americans. Other islands and battles in the middle of nowhere, or so it must have seemed to the men who struggled and perished there, are islands and places most people have never heard of: The Aitape-Wewak Campaign at Sepik River, Ambon Island in the Banda Sea, Arawe, and battles named Enogai, Kaiapit, Munda Point, Piva Forks, or The Battle of the Green Islands in Papua New Guinea. Others have need of spell-check: the Battle of Komandorski Islands, the Finisterre Ridge Campaign, Masbate (need to watch the pronunciation on that one), the Landings at Cape Torokina, and Operation Tan No. 2 (not sure what happened to Operation Tan No. 1).

The ocean depths hide fallen warriors forever and seldom does a Pacific jungle give up the missing. Unmarked graves, a rusting tank in an unexplored jungle, downed airplanes enveloped by vegetation, an old photo, a bone, a rifle, a boot, all can give a clue to a fallen soldier or airman in the middle of nowhere.

A rainforest or mountainous terrain can also conceal the living. Five months after Japan's surrender, 120 Japanese soldiers were routed in a mountainous battle 150 miles south of Manila. On Guam the same year, abandoned Japanese soldiers killed a six-man American patrol.

During March and April of 1947, Japanese Lt. Ei Yamaguchi led 33 soldiers in renewed attacks on American Marines stationed on Peleliu Island. A Japanese Admiral eventually convinced the holdouts that the war had been over for two years.

On June 30, 1951, about 30 shipwrecked Japanese (and one Okinawan woman) on Anatahan Island, 75 nautical miles north of Saipan in the middle of nowhere, were at long last convinced to surrender. During their prolonged isolation, at least six of the males had been killed fighting for the female's attention. They had survived on coconuts, taro, fish, lizards, and wild sugar cane. They smoked dried papaya leaves wrapped in banana leaves and consumed "tuba" (coconut wine). If not fighting for the woman, they fought when intoxicated.

Tinian: The B-29 Enola Gay took off from this island airfield in August of 1945 to drop the atomic bomb on Hiroshima, Japan. A Japanese soldier, Marata Susuma, was finally captured in his tiny shack near a Tinian swamp. The year was 1953.

January, 1972: On the banks of the Talofofo River, soldier Shoichi Yokoi is finally found, still carrying his army-issued knife.

Lubang Island, 1974: 2nd Lt. Hiroo Onada surrenders 29 years after the end of WWII. Onada had been declared legally dead in 1959.

Soldiers Kiyoaki Tanaka and Shigeyuki Hashimoto, who had fought alongside insurgents of the Malaysian Communist Party, finally surrendered in Thailand in 1968.

In 1997 and 2005, there were persistent reports of Japanese soldiers on Mindanao and Vella Lavella, 50 to 60 years after the war ended.

Following Pearl Harbor, the first island battle was on a small Pacific Atoll called Wake Island. The American defenders, cut off from resupply and reinforcements, would fight an amazing defense against unbelievable odds, way out there in the middle of nowhere.

HENRY TALMAGE ELROD
"HAMMERIN' HANK"

Wake Island is a pint-sized coral atoll in the middle of nowhere, 2,300 miles west of Honolulu and 1,510 miles east of Guam. This tiny speck of sand and palm trees consists of three islands, Wake, Wiles, and Peale, with a combined shoreline of 12 miles. The highest elevation is 20 feet. History would record Wake Island as the only battle in WWII where an amphibious assault failed when a ragtag group of American Marines, sailors, civilian workers, and 45 Chamorro Islanders turned back a Japanese invasion.

Turner County, GA was established on August 18, 1905 with Ashburn as its county seat. Northeast of Ashburn, pretty much in the middle of nowhere, is a pint-sized town called Rebecca. The city's land mass is 0.8 of a mile. In the year 2000, a census recorded the population at 246. A little more than a month after the founding of Turner County, Henry Talmage Elrod was born in Rebecca, GA on September 27, 1905—smack dab in the middle of nowhere.

Fate would fuse the small-town Georgia boy to an equally tiny island in the North Pacific Ocean. During the Battle for Wake Island, Elrod would engage 22 enemy aircraft singlehandedly, be the first American to destroy a major Japanese Naval vessel from a fighter aircraft, assume command of soldiers on one flank of the ground defenses, and be the first American in WWII to receive the Medal of Honor. He was also the first Marine pilot Medal of Honor recipient to be married to a Marine.

Elrod attended the University of Georgia and Yale University before enlisting in the Marine Corps in December, 1927. He earned the butter bars of a 2nd Lieutenant in February, 1931. After over a year at the Marine Corps Basic School in Philadelphia, student aviator Elrod reported to the Naval Air Station in Pensacola, FL where he pinned on his wings in February, 1935.

Henry Elrod, MOH Recipent

Transferred to Quantico, Elrod served as an aviator plus pulled duty as Officer in Charge of his squadron's welfare, training, and personnel. Transferred to San Diego in 1938, he served in a variety of roles until dispatched to the Hawaiian area in January, 1941.

December 4, 1941—three days before the Japanese attack on Pearl Harbor, Elrod, now a captain with fighter squadron VMF-211, flew one of 12 F4F Wildcats to Wake Island. Since Wake Island is on the opposite side of the International Date Line, history records December 8th as the date the Japanese attacked the island, only hours after their assault on Pearl Harbor.

During the first air strike, Japanese bombers from the Marshall Islands demolished eight of 12 F4F Wildcats on the ground. The remaining four Wildcats were in the air on patrol, but failed to see the enemy bombers due to limited visibility.

December 11—the Japanese attempt an invasion of Wake Island using three light cruisers, six destroyers, two patrol boats, and two transports with 450 special Naval landing troops. Against this armada stood 449 US Marines, including pilots, 68 Navy personnel, five Army, over a thousand civilians, and 45 Chamorro Islanders. The defenders waited until the Japanese ships were well within range. When they did fire, salvos from a shore battery scored a direct hit on the Japanese destroyer *Hayate's* ammo magazine. The *Hayate* exploded and sank within two minutes, carrying down her entire crew of 167 men. Shore batteries scored damaging hits on three more Japanese warships.

In the skies, the four remaining Wildcats of VMF-211 were on the prowl. Capt. Elrod spotted 22 incoming Japanese aircraft and engaged them by his lonesome. He shot down two. Then Elrod participated in strafing runs and low-altitude bombing against Japanese shipping. Another Marine fighter pilot confirmed that at least one of Elrod's 100 lb. small caliber bombs hit the Japanese destroyer *Kisaragi.* Elrod placed the bomb in the perfect place; right in the middle of an extra load of depth charges. The same Wildcat pilot later saw the *Kisaragi* blow up. Like the *Hayate,* the *Kisaragi* carried all 167 of her crew to a watery grave.

Japanese destroy the Kisaragi, sunk with all hands

Capt. Elrod limped back to base with a perforated oil line and other battle damage from anti-aircraft fire. He landed safely, but his Wildcat was totaled. Elrod became a grunt, fighting on the ground in command of one flank of the defending forces.

The dishonored Japanese fleet commander reported they had been "humbled by sizeable casualties" then quickly retired to the Marshall Islands to lick his wounds. Humiliated, infuriated, and seeking retribution, the Japanese returned on Dec 23rd with many of the same ships plus two aircraft carriers, two battleships, and

2500 additional troops. Overwhelmed by the firepower and sheer numbers of the invasion force, the Wake defenders surrendered after over twelve hours of fierce combat, much of it hand to hand.

Captain Henry Elrod conducted an exceptional land defense in his sector. He and his men repelled a number of attacks while providing cover fire for unarmed ammunition carriers. During one attack, Elrod captured an enemy automatic weapon, gave his own weapon to another man, then both men fought on as best they could. As he continued to provide cover fire for his ammunition carriers, Elrod was hit and fell mortally wounded.

The Georgia boy from the middle of nowhere fought and died for his county in the middle of nowhere.

Their stubborn yet courageous defense of Wake Island cost the Americans 120 killed, 49 wounded, two missing, 12 aircraft destroyed, and internment for the 433 surviving military personnel and over 1100 civilians. Eventually, five of the military POWs would be executed.

The Japanese paid heavily for this tiny speck of sand: 820 killed, over 300 wounded, a cruiser seriously damaged, two destroyers sunk by Captain Elrod, two transports sunk, one submarine sunk by an American sub, over a dozen aircraft lost and another 10 damaged, plus a four-year occupation of an island in the middle of nowhere.

On November 8, 1946, Elrod's widow was presented the Medal of Honor by President Truman for her husband's courageous defense of Wake Island, posthumously. Elrod was also promoted to the rank of Major. Initially buried on Wake Island, in October of 1947 Major Henry Elrod was brought home and reinterred in Arlington National Cemetery.

"War is a series of catastrophes which result in victory."
—Georges Clemenceau

FLYING FABRIC AND PILOTING PLYWOOD

Anti-aircraft tracers swivel and coil around your aircraft like a deadly viper; the flak is thick as molasses, and you and your passengers are going down. The normal descent is 72mph, landing speed around 60mph, but if you miscalculate and drop the speed to 49mph your aircraft will stall and crash, killing everyone aboard. Your flimsy flying machine was produced by a variety of recognized American manufacturers, including beer conglomerates Anheuser-Busch and Schlitz Brewing Company, Gibson Refrigerator, the Ford Motor Company, Ward Furniture, a piano manufacturer, even a coffin company. You are sitting behind the rudimentary instruments and controls of a Waco CG-4A Glider.

The air armada is 1400 aircraft strong and taking heavy enemy fire, yet the plywood pilot's seat is so uncomfortable on long missions you utilize your parachute as a seat cushion. Shrapnel rips through the fragile fabric overlaying a metal airframe as other projectiles punch holes in the plywood floor. Motorized power is non-existent since the aircraft was constructed without an engine and as you slip to earth the soldiers onboard pray you paid attention during pilot training because they know landing in a glider only results in two outcomes: very good or very dead.

In late 1941 and early into the war, the Army utilized sailplanes for glider training. That decision was a whopping blunder since sailplanes take advantage of thermal conditions and can soar for hours with a skilled pilot behind the controls. Not so with Army glider pilots. Their abilities required aerial truck drivers with the proficiency to land the equivalent of a trailer with a pair of wings in combat conditions. A casualty rate of 78% was not unusual. Most glider pilots only flew one or two missions; Guy Gunter flew four major combat missions, and this is his story.

Atlanta resident Guy Gunter was born on June 6, 1918. He attended Murphy High School and Tech High and recalled the day Pearl Harbor became a household word. "I was employed by General Electric at the time as a traveling salesman and was having lunch in an Atlanta restaurant when I heard the news about the Pearl Harbor attack. You know, every individual in that restaurant knew a war had started but I didn't think too much about it. I guess I was too young to understand the significance."

Guy preferred aviation. In late 1942, he interviewed with a Navy Flight Board in Macon and was accepted for pilot training. "That's what I wanted to do," he said. "But in the meantime, I received my greetings telegram from Uncle Sam. At that time in the war, a draftee could go into the Army and still receive a transfer to the Navy when a slot opened in pilot training. So, the Army sent me to Shepard Field in Wichita, TX for instrument training then

I became an instructor and waited for a call from the Navy. And I waited, and I waited. That call never came because the transfer program had been terminated."

Guy applied for and was accepted for OTS (officer training school). "I made friends with one of the instructors," he said. "I told him I wanted to fly. He called me two days later and offered me glider school. I took the deal." Thus began Guy's lifelong love with aviation.

Sent to Hayes, Kansas, Guy learned to fly from fields cut out by tractors and bulldozers. "I got my wings in Piper Cubs," he said. "I soloed after four hours of training. I loved it. I kept doing a 'touch and go', you know, practice landings and takeoffs. An instructor finally had to wave me down. Shoot, I didn't want to stop."

Advanced glider training took place at Lubbock, TX. Guy recalled, "Actually, what they considered a glider was an airplane with the engine removed. We trained with a couple of different aircraft, an Aeronca and a Taylorcraft. Without engines, there was room for three seats, trainee in front, instructor by his side, and a second trainee in the back. When the Waco gliders finally arrived, they were pulled in by L-2s with fighter jocks piloting the gliders. It was funny to watch the fighter pilots. They were scared to death wrestling with dead stick aircraft."

The first class of glider pilots graduated on November 14, 1942. "There were 109 of us in the class," Guy recalled. "The Army said we would be assigned as instructors in Austin, TX. Yeah, right. We found ourselves on the former luxury liner SS Maripose headed for Egypt by way of Rio de Janeiro, an unescorted 45-day voyage."

The barren land of Egypt was also barren of gliders. Guy recalled, "I flew copilot on C-47s or C-46s for about 30 days. We were also relocated to Algiers and Libya on occasion to fly cargo flights into Gibraltar, Oran, and North Africa. Of course, we also flew secret whiskey runs back to Egypt."

Gliders appeared overnight. Guy stated, "I flew into Algiers one morning and saw gliders all over the place. Our training began immediately, very intense training. We had a lot of fun but all of us knew something big was up." That 'something big' was the invasion of Sicily.

C-47s releasing gliders over Normandy

Guy said, "We trained the British pilots for three weeks then they asked for American volunteers. Well, being as foolish as I am,

I volunteered. We had 35 glider pilots on the first flight; only 16 of us made it back. It was a real slipshod affair. An old Army sergeant piloted our tow plane. The thick anti-aircraft fire got on his last nerve so he wanted to cut us loose. I knew we were too far out from the port of Syracuse and told him so; I said we couldn't make dry land. Obviously, the man didn't care.

Here comes the rope; he'd cut us loose. We flew into a hornet's nest. Shrapnel peppered my face and legs, plus the glider took a lot of damage but the flight characteristics were still okay. I put her down in the bay. We stayed in the water all night clinging to the glider. A Captain in our waterlogged group told me to say a prayer for everybody. Shoot, I told him to pray for himself, I was too busy praying for yours truly!" Luckily, the next day a Greek destroyer picked up the men, instead of German Naval vessels.

Transferred to Sicily, Guy flew cargo planes until assigned as a pilot on a four-seat single engine Fairchild 24. "I flew the big brass around the island for about six months," he said. "As you may have noticed, there's plenty of spare time between missions for glider pilots. Shoot, I had a great time, but it didn't last long." Guy received another transfer, this time to England to prepare for the Invasion of Normandy: D-Day. Asked his opinion of the English people, he replied, "Well, at least the girls understood what the heck we were saying."

June 6, 1944: On Guy's 26th birthday, General Eisenhower sends the Allied armada across the English Channel to invade Europe. Guy recalled the Normandy Invasion: "We took off just after midnight, carrying a pathfinder group of the 82nd Airborne. An hour later, we put down about 25 miles south of the main front. We had no problems, we made it down, we were very lucky. I brought the glider in nose high, we hit on the tail and plowed straight into a hedgerow. That was okay though, it stopped the glider. We'd taken Dramamine pills and a shot of scotch so we were up to the task. I remained on the ground for about six days." Asked what glider pilots do after landing, Guy replied, "Try to stay alive. Most of us had a carbine or Tommy gun so we joined the fight."

Guy's next port-of-call: a small village near Mount Vesuvius in Italy. "There wasn't much to it," he recalled. "The runways were bulldozed dirt and not much more." Operation Dragoon was on the horizon; the men and gliders made ready for the invasion of Southern France.

An Army glider, and the cost of freedom

Guy said, "The fields in southern France were not much of a problem. The real problem was all the tall poles the Germans put up to prevent glider landings. Shoot, we didn't care, we landed between the poles. The poles tore the wings off but that kept the fuselages on a straight course so we did okay. The paratroopers had a rough go of it. As they oscillated in their chutes, a lot of the guys would slam into the poles, a lot of those boys broke their backs."

The charcoal burning car: "We fought against the Germans and Vichy French for about 48 hours. I didn't have a lot of love for the Vichy French, none of us did. Anyway, after things calmed down, I 'confiscated' a 1936 Model V8 Ford with a huge charcoal burning tank on back. We had a blast in that car until we had to quit." When asked if 'quit' meant going back into action, Guy responded, "No, the dang fire burned out."

Guy returned to England to participate in the largest airborne assault in military history: Operation Market Garden, the failed attempt by British Field Marshall Montgomery to enter Germany via the Netherlands. Airborne resources flew 34,600 troops into combat; 20,011 by parachute, 14,589 in 3,140 various gliders. A shortage of American and Allied glider pilots meant using one pilot per glider with an additional soldier occupying the copilot's seat. The glider air armada was pulled to their landing assault areas by 1,438 C-47 or C-46 cargo or transport aircraft. These statistics do not include ground force participation in Operation Market Garden.

Guy recalled, "The gliders in our formation flew directly over a German anti-aircraft school. A bit scary when you think about it, but we came in too low for them to hit anything. We had a real smooth landing and we all got out okay, if 45 miles behind enemy lines can be considered okay. I grabbed my weapon and joined the combat, just trying to stay alive, as always. We assaulted the bridge at Nijmegen."

Pausing a moment, Guy continued, "When a glider is cut loose from the tow plane, the pilot wants to get down as soon as possible because they are always shooting at you." When asked to explain the meaning of 'as soon as possible', Guy responded, "As fast as humanly possible!" One glider flight characteristic seemed unnerving, especially to yours truly, an amateurish pilot who fancies an engine attached to his aircraft. Guy recalled, "We never looped a glider on a combat mission, but I did loop several gliders in training. However, we couldn't roll one." When asked why not, Guy replied, "Let's put it this way, I never tried."

Against all the odds of combat, Guy Gunter survived his fourth major aerial assault as a glider pilot. He returned to England and flew as copilot on cargo planes until assigned to Reims, France to pilot C-46 Commandos. Restless and bored, he volunteered for the last glider mission of the war into the heartland of Germany, but the Army denied his request. "They said four glider landings in a combat situation was enough," Guy recalled. "I guess the Army was right for once. Two of my best friends at the time flew into Germany; it was their initial mission as glider pilots. Neither one made it back."

After the war, Guy eventually went into his own business as an appliance distributor and stayed in the air with his personal aircraft, a Beech Bonanza and twin-engine Beech Baron. He still remembers one flight to Florida. "Shoot, the Bonanza swallowed a valve and blew a hole in the block. I cut air to the engine which put the fire out, but oil was all over the windshield and I had no power. So, I put'er down, took six trees with me and landed nose up. I had chipped bones in my ankles and the guy with me broke his pelvis, but nobody really got hurt. Lost a good plane, though."

He also landed the twin-engine Baron without an operable nose wheel. "No big deal," Guy said. "I kept the nose up until low airspeed and gravity forced the nose and twin props into the runway." Guy's hours behind the controls of military and civilian aircraft, including over 1,800 hours in gliders, approached 30,000 when he finally retired his wings at a ripe young age of 92.

Asked if he missed the wild blue yonder, Guy replied, "I still have 20/20 vision, my faculties, and this dang walker. Tell you what, take me to an airport and stand me next to the wing. I'll crawl up into the cockpit, I'll pull myself into the pilot's seat, and I will get that baby airborne. You bet'cha."

I interviewed Guy Gunter several years ago. As cocky as all aviators are known to be, the talk we shared was one of the most pleasurable of my writing career. Guy passed from this life on November 2, 2016 at the age of 98. No debate is required concerning his position beyond the Pearly Gates, Guy earned his wings a long time ago.

THE WARRIOR'S GRAPEVINE

I Heard it Through the Grapevine by Marvin Gaye is one of my favorite melodies. I enjoy the rhythm, Gaye's smooth vocal sound, and the haunting truism of human conduct with the insinuation of a Motel 6 liaison as described by music. The title, however, can easily be applied to the *Warrior's Grapevine* used by veterans to track down and hopefully help a troubled brother or sister from a life of mental isolation.

The *Warrior's Grapevine* abides in the hearts of veterans that are willing to reach out for their comrades existing in a world outside of actuality. Veterans able to successfully transition back into civilian life understand that too many of their brothers and sisters discard their uniforms but their souls remain on the battlefield. Basically, the distressed veteran is faking readjustment.

Military unit reunions and/or service magazines may offer temporary reprieves from the undisciplined dog eat dog world, and organizations like the Veterans of Foreign War and American Legion provide welcoming atmospheres to share a beer and a war story with comrades more likely to understand the pain. But participating in a bullshit session with battlefield buddies is not the answer.

Without professional help, many veterans continue to suffer in silence by blocking out or bottling up unpleasant memories. If employed, they battle the daily rat race. In slumber, they relive battles from long ago. Often just a single incident from combat can scar their psyches for life. Trench Disease, Shell Shock, Combat Fatigue, or PTSD, call it what you may, war changes a warrior forever. Most veterans make the readjustment to civilian life relatively stress-free, while others do not. But no veteran can avoid the transformation from warrior back to civilian. Most recruits go into the military full of piss and vinegar and filled with romanticism begotten of patriotism. Then they are taught to break the sixth commandment lest their own ass be shot off. Many come home broken in spirit and drained of all innocence, never again the same kid that signed the dotted line.

The *Warrior's Grapevine* achieves its understanding and dynamism from experiences in war, and for those that do transition successfully, their sense of duty remains the same: don't leave anyone behind.

I heard it through the *Warrior's Grapevine* that a veteran of Iwo Jima lived just north of Monticello, GA. Scuttlebutt had it that he'd been on top of Mount Suribachi guarding the famous flag-raisers with his BAR (Browning Automatic Rifle) and that he was one of the few Marines of Easy Company to make it off Sulfur Island alive. I gave him a call and set up an interview.

Traveling from Atlanta out I-20 East toward Augusta, I turned right at the Social Circle exit and followed the whiff of cow manure. Easy to do with dozens of cow pastures and thousands of cows flanking the highway. I passed a flat-top boulder called Preacher's Rock, so named for wagon train preachers who stood atop the big rock to spit fire and brimstone down upon immoral trail blazers. About three miles down the road, I made a pit stop at a country-as-corn convenience store that sported one workable gas pump and an inoperable toilet. Inside, the store was reasonably hygienic, considering the outside ambiance.

A short distance from the convenience store from hell, the county highway gave way to a gravel road riddled with potholes from hell. Hell, I bounced and bumped yet still managed to keep my mouth open so as to avoid chipping any teeth. After about a mile of bone-jarring choreography, I took a sharp right turn onto the veteran's property. His rock-strewn driveway was in better condition than the gravel road and meandered about a quarter mile to a pleasant-looking ranch style house with a workshop out back the size of Trump Tower. A small pond was nearby, the water low, a couple of ducks bitch-quacking due to my early morning intrusion. A cluster of grass-fed range-roaming chickens clucked and fussed as I exited my car. As I side-stepped the aftermath of fussy chickens, I heard a rumbling deep-gutted growl.

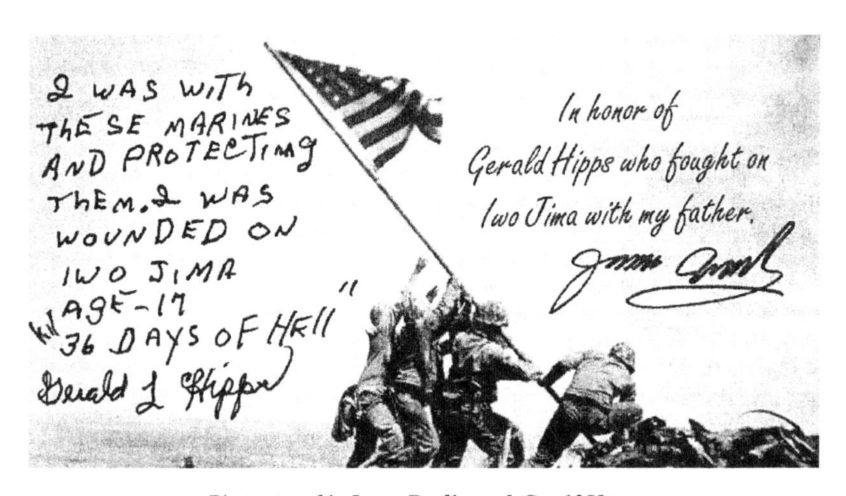

Photo signed by James Bradley and Gerald Hipps

A pallid Pitbull with a head the size of a basketball sat on his haunches about two feet in front of me. Its loud rumbling faded into a judicious grumble, as if debating a welcome or a funeral.

"Don't worry 'bout him," someone shouted from the back porch. "He ain't bit nobody in almost a year."

The less-than reassuring remark came from Gerald Hipps. A willowy fellow, thinning hair, most likely in his late eighties but appearing to be in pretty good shape. Not a tooth in the man's head, but there was a darn good reason for the lack of choppers.

"Are you Mister Hipps?" I asked.

"Sure am," he replied. "Com'on in."

"Uhhh, what about the dog, sir?"

"Bring him in with ya."

Mrs. June Hipps, a sweet lady with a hospitable manner, joined me, the Pitbull, and Gerald for the interview. He had not spoken of Iwo Jima until recently. He explained why, "My son and his wife noticed a pickup truck with the bumper sticker 'Iwo Jima Veteran', and I thought, 'Well, shoot, how come I'm keepin' my big mouth shut.' So I decided to start talkin'."

Gerald Hipps, USMC

Only in the last couple of years had he made contact with the nearest field director for the Georgia Department of Veterans Service and agreed to visit the VA Hospital in Atlanta for proper evaluation. Hipps was eventually given disability payments for PTSD. "I was invited to speak at the VFW in Covington," he said.

"Don't remember much about it, but I guess I did okay. It was a little nerve-rackin' for me."

"I'm glad you're speaking out," I replied.

"Well, I plan to keep on doin' so," he said. "How'd ya find out 'bout me?"

"We call it the *Warrior's Grapevine.*"

"What's that mean?"

"It means veterans helping other veterans with reverence and empathy without bureaucratic busybodies sticking their noses in our business."

"I like that," he replied, nodding his head. "Well, sir, ya ready to do this?"

"Yes, sir, you go right ahead."

Before he began in earnest, Hipps requested, "I'd greatly appreciate it if you wouldn't interrupt me while I'm talkin'. I have this thing sorta memorized. It's the only way I can tell it, if that's okay with you."

That was fine with me. Besides, I needed to keep an eye on the Pitbull.

THE MARINE FROM MIAMI

Born into extreme poverty near Miami Beach, Gerald Hipps and his family lived without electricity or indoor plumbing. Steady meals were rare; his mother worked in a laundry; a grandmother raised the siblings; a strawberry was a special treat, as was his first pair of shoes at age 10.

Hipps said, "I was at the movies with a friend when we heard the news about Pearl Harbor, so I decided to join the Marines." He did so with his mother's written permission: Hipps was 16 years old. With a reporting date three months away, Hipps found interim employment building PT Boats for the Navy. "I needed to make money for the family," he said. "But that job didn't last too long because I got fired." It seems another worker called Hipps a dirty name. "I sorta beat him up," Hipps said with a smile.

Although physically tough, Hipps discipline had much to be desired during basic training at Parris Island. "I was too cocky," he recalled. "A drill instructor told us to 'shut-up' and not say nothin' while he taught us about weapons but I asked him, 'Well, can we smoke?' so he got a little angry." Ordered to put a string around his waist, pick up cigarette butts then tie the butts around the string, Hipps told the DI, 'Man, you must be crazy.' In due course, Hipps obeyed the order, plus had to do a Hula dance in the process. "I guess that should have taught me a lesson," he confessed. It hadn't.

Surviving basic training, to some extent, Hipps asked for training as a cook. He stated, "I was sick and tired of being hungry all the time so learnin' how to cook made a lot of sense to me." His cooking career lasted less than two weeks. "The mess sergeant caught me workin' on my curve ball," he said, smiling. Apparently the Marines do not appreciate rookie cooks who practice baseball by tossing Uncle Sam's peeled potatoes against the mess hall wall. Hipps was sent straight to advanced weapons training.

"I became a BAR (Browning Automatic Rifle) man," he said. Still too undisciplined by Marine standards, Hipps' impetuous behavior culminated in one final punishment: perched atop exposed ceiling beams in the barracks, naked as a Jaybird, with weapon in hand and singing loudly, '*I'm a gooney bird from Buford.*' He confessed, "I sorta caught on to the discipline thing after that."

Trained, tough, and reasonably disciplined, Hipps' next port-of-call was Camp Pendleton, CA for more training before boarding a ship in San Diego. The ship joined a massive convoy of aircraft carriers, battleships, cruisers, destroyers, troop ships, and support vessels zig-zagging across the Pacific Ocean to evade enemy submarines.

Once at sea, Company E, 2nd Battalion, 28th Marines was ordered below deck to learn their destination. Hipps said, "We gathered around this long table with a mock-up of an ugly little island with an uglier large hill that we had been ordered to take." The ugly little island was Iwo Jima; the uglier large hill Mount Suribachi.

February 19, 1945 at 0400—A breakfast of steak and eggs. 'Eat up,' the Marines were told. 'It may be your last meal.' The Japanese propaganda queen, Tokyo Rose, transmitted dismal warnings to the teenagers boarding the LSTs: 'You sailors and Marines on Iwo Jima will die while your wives and girlfriends cheat on you back home. Give up. The Marines are already pushed back into the sea.' Hipps said, "Shoot, we weren't even off the ships yet and she was sayin' we'd already been whipped. She played some pretty good American music though."

Hipps and Easy Company climbed down rope ladders into

landing craft, cautious to keep their hands on the vertical ropes, not the horizontal, or the man above would step on and break your fingers. The landing crafts began their deadly gauntlet for the black sulfur sands of Iwo Jima. Enemy shells missed landing crafts; scored direct hits on others. Boats and Marines disappeared. Hipps recalled, "That was the really scary part. We kept our heads down but those Navy guys, well, they had to keep their heads up because they were drivin' the boats. They were very brave men."

Hipps made the beaches with Easy Company. He recalled the scene vividly. "The beach was terrible" he said. "Dead Marines were all over the place and wounded Marines were screaming for a corpsman. It was mass confusion. I heard guys screamin', 'Get off the beach, move, move, get off the beach!' that's about the time I got hit by shrapnel."

Wounded and bleeding, Hipps yelled for a corpsman. The hot shrapnel had ripped flesh on his legs, arms, neck, and shoulders. "A corpsman found me almost immediately," Hipps said. "He put sulfa on the wounds and patched me up pretty good." The corpsman was John Bradley, one of the six famous flag-raisers and the father of James Bradley, author of the world-acclaimed book and movie directed by Clint Eastwood, *Flags of our Fathers.* Treated, rested, and determined to rejoin Easy Company, Hipps refused medical evacuation then left the death and suffering on the beach to enter the hell called Iwo Jima.

Lost and separated from Easy Company, Hipps skirmished foxhole to foxhole until reunited with his band of brothers scaling Mount Suribachi. "That's when I finally found big old Ira," Hipps said. "I was always behind Ira, even in the chow lines back in the States. He was a great Marine, but a trouble-maker and hell-raiser if he'd been drinkin' a bit." American Pima Indian Ira Hayes was one of the celebrated flag-raisers on Mount Suribachi. Born and raised on the Gila River Indian Reservation in Arizona, he survived the demons of Iwo Jima but lost the battle with his own demons after an all-night game of poker and binge-drinking on Jan. 24, 1955. He was 32 years old.

Hipps and Easy Company took the summit of Suribachi to witness and/or participate in both flag-raisings. "I watched both the photos taken," Hipps recalled. "The first flag was real small, but when the Marines on the island and all those Navy ships saw the flag all hell broke loose with whistles and cheers and ship horns a blowin'. It was great."

The second larger and most-recognized flag-raising was virtually a non-episode on Iwo Jima; no bells and whistles, and no ship horns. Award-winning combat photographer Joseph Rosenthal almost missed the celebrated shot, swinging his camera around at the last second to click off a frame. He didn't even take time to glimpse the image in the viewfinder, yet that 1/400 of a second camera speed snapped the world's most famous combat photo: The six Marines raising Old Glory on Iwo Jima. And where was Gerald Hipps? He said, "I was standing in the background protecting the guys with my BAR."

Easy Company was ordered off Mount Suribachi after the flag-raisings. They were told their war was over; they were going home. Hipps said, "We got to the beach and found out things had changed. They turned us around and pointed toward the other end of the island." Easy Company wasn't going home; they were entering the barren hell of Iwo.

Hipps was in brutal combat at close range, many struggles hand-to-hand. He recalled, "I was fightin' all the time, a cruel way of fightin', kill or get killed. The north end of the island was nothing but a killing field. One Jap was hidin' beneath a Marine's rain poncho and just jumped up and killed my buddy standing next to me. We killed the Jap, but I was a lot closer to that Jap than my buddy was. I've always wondered why he didn't shoot me."

In another incident, Hipps was ordered to rescue a Marine trapped at the bottom of a ravine. "I tried but got pinned down by sniper fire. We called up the flame-throwers to handle the situation." To the Marines, a foxhole was home. "We slept two to a hole," Hipps said. "You'd sleep for four hours then be on watch for four hours. But nobody got any sleep because out of the blue a Jap would jump in your hole and you'd be fightin' for your life."

Daily progress was measured in yards, 100 or 200 yards, many days even less. Hipps recalled, "It was like yesterday; our colonel called some of us into a group, maybe about 20 of us, for a quick briefin'. Well, a Jap grenade dropped right in the middle of us and exploded. Most of the guys got killed but I didn't get a scratch. I was shakin' like a leaf but the colonel said, 'Don't be afraid, son,' and I remember those words to this day. I loved that guy. He was killed later that afternoon."

Friends were made, friends were lost. "My best buddy, a fellow named Marino, was sharin' a foxhole with me," Hipps said. "He was quite a character, a gamblin' man, kept loaded dice and a deck of marked cards in his pockets. I really loved the guy, but he stuck his head up to see what was goin' on and, well, got a bullet right through his head. Those kinda things are tough to forget."

The Marines kept going, kept dying, kept fighting, kept their 'esprit de corps' and kept on winning. Hipps stepped on a landmine. "I knew I was a goner," he said. "It clicked and I waited to meet my maker, but it turned out to be a dud." Hipps and the remnants of Easy Company made it to the northern end of the island. He said, "I knew the fightin' was almost over, but I had this horrible feelin' my luck was runnin' out. I prayed, 'God, please let me go home,' and I still pray to him every day."

One hour after his short prayer, Easy Company was relieved by the United States Army. "What was left of us couldn't believe our good luck, we were still alive," he said. Easy Company initiated a calculated withdrawal to what should have been a secured airfield. "Nothin' was secure on Iwo," he said. "A bunch of Japs had slipped around us and got to the airfield before we did. Air Force boys were walkin' around the airstrip because B-29 bombers and fighter planes were already makin' emergency landings on Iwo. The Japs attacked and killed all those boys." Even after the word "secured" entered military history on Iwo Jima, several pilots and crews were butchered in their tents by roving bands of Japanese survivors. Of the probable 21,000 Japanese soldiers on Iwo Jima only 216 were captured or surrendered. The last two Imperial soldiers, Matsudo Linsoki and Yamakage Kufuku, crawled out of a concealed cave and surrendered in 1951.

Hipps talked about the flag-raisers. "Of the six, only three got off the island alive. Old Ira made it, and John Bradley and Rene Gagnon. Harlon Block, Mike Strank, and Frank Sousley were killed on the island, but because of the photo they will live forever in Marine Corps history." Almost 6,000 Marines lost their lives, 17,000 were wounded, yet the lives of approximately 25,000 American aircrews were saved due to a 'secured' Iwo Jima. The young Marines did not die in vain. Gerald Hipps landed on Iwo Jima with 240 of his buddies. He was one of 27 that made it off the island.

What remained of Easy Company and Hipps boarded a ship and set sail for home. 'Home' meant the Hawaiian Islands to prepare for the upcoming invasion of Japan. Hipps said of the voyage, "I had nightmares on the ship and I still had my BAR with me. One night a Marine ran through the door yellin' from his own nightmares and scared all of us awake. The first thing I saw was the red exit light above the door. In my confused state of mind, I recognized the red light as ordnance headin' in our direction. I grabbed my BAR and almost shot that screamin' young man; thank the Lord I came to my senses before I pulled the trigger."

The two atomic bombs nullified the dreaded invasion of Japan. Instead of another invasion, Hipps received orders for Sasebo, Japan to help gather Japanese weapons and search houses for war materials. He said, "The Japanese showed me the utmost respect when I had to search their homes. Several asked me to stay for dinner. The tomatoes were the prettiest I'd ever seen, and very delicious, but had I known their vegetables were fertilized by human manure I would have politely refused to eat them."

Returning home, Hipps received his honorable discharge in San Diego. The government offered him the equivalent of a year's salary as part of a severance package. Hipps said, "I refused the money. I didn't want a dang thing from the government. Iwo Jima and what I witnessed on that island had changed me forever. I bought some civilian clothes and walked over to a YMCA, changed clothes, and left my uniform in the middle of the floor. My war was over, my duty was done, and I was done with it." Gerald Hipps was 19 years old.

Hipps returned to his mother and grandmother a hardened warrior, with permanent scars on his body and gruesome images of Sulfur Island implanted in his memory forever. For the rest of his life, he would fight and endure the battles on Iwo in vivid nightmares and cold sweats. His family knew nothing of his ordeal; his future wife never saw his Marine photo, his children and grandchildren and great-grandchildren had never heard his story.

The self-imposed silence was only broken a few years ago. Hipps said, "Ya know, when my son and his wife saw that old fogey drivin' his pickup with a bumper sticker 'Iwo Jima Veteran' that's when I thought, 'Shoot, I can do that' but I couldn't find a bumper sticker. So I went down to Walmart and bought a pack of stencils then painted on my own bumper sticker. I reckon it did the trick 'cause people started callin' me and askin' me questions about my service. I didn't think anybody cared."

At the age of 85, Gerald Hipps received a call from writer and author, James Bradley, son of the Navy corpsman and flag-raiser, John Bradley. "He asked me to join him in Atlanta for a speech and book signin' event," Hipps said. Hipps was picked up and transported to the event by a United States Marine. "I sat in the audience and listened to Mister Bradley speak. I was really inspired by his remarks. When he asked for questions, I raised my hand and told him, 'I served on Iwo Jima with your father'." Mister Bradley replied, 'You must be Mister Hipps. Please, come join me on the stage.' I couldn't believe it. As I walked to the stage, the folks gave me a standing ovation. I'd never been applauded in my life. Some of them even wanted my autograph, can you believe that?"

Gerald Hipps is at long last speaking out and articulating his story. With help from the Georgia Department of Veterans Service, he's received medical and psychological help from the VA. He's even given a speech. "I wasn't too good at it," he suggested. "I rattled on for the better part of two hours, so maybe I did need to talk."

Retired from the carpentry and construction business, Hipps takes pleasure in the company of friends and family. "Two of my boys fought in Vietnam," he stated. "They weren't treated very well by folks when they returned from that war, so it's time for them to speak up, too. We all have a story to tell."

*****

After the interview, Hipps and the Pitbull escorted me to my car.

I had made a lifelong friend with Gerald Hipps, but I still wasn't too sure about the dog.

We made small talk as the moody ducks continued to quack their displeasure from the safety of the small pond. Hipps shook my hand and said, "I appreciate ya comin' out to talk to me."

"It was my honor, sir," I replied.

The feeble buzzing of trail bikes and four-wheelers behind a stand of pine trees broke our attention until the buzzing became an ear-splitting roar, then gradually faded out into a distant tree line.

Hipps stated, "I reckon the kids are bikin' today."

"Aren't they on your property?" I asked.

"Yeah, but I don't mind," Hipps said. "All the youngins' around here bike the back forty every so often; they're just havin' a little fun. All my grandkids bike, too, that's why I haven't bought myself some new teeth. I saved over three grand to buy some teeth but the grandkids wanted a four-wheeler. Shoot, I'll buy some teeth sooner or later but my grandkids won't be youngsters forever. I like to see them smile."

* * * * *

August 21, 2012: Gerald Hipps was caring for his beloved wife, June, recently released from a hospital. His morning chores completed, Hipps told June, "I'm goin' lie down on the couch for a bit."

Hipps laid down, closed his eyes, and slipped gently away. Another Leatherneck had gone home.

Hipps lacked the documentation to prove he was wounded on Iwo Jima. Denied benefits for his injures, even with the support of famed author James Bradley, Hipps' wife, June, has now refiled: It seems after the cremation shrapnel was found in Gerald's ashes.

DEATH MARCH, A HELL SHIP, AND A COAL MINE

The Philippines, April 9, 1942: Roughly 60,000 to 80,000 Filipino and American POWs begin the forceful relocation from Saisaih Point and Mariveles to infamous Camp O'Donnell, a trek of 60 miles. The trudge soon became a murder binge, costing the lives of untold thousands of soldiers who perished from beatings, malaria, dysentery, beheadings, and bayoneting. The death estimates vary due to random killings, mutilations, no burials, and tank treads churning fallen men into pulp. History records this butchery as The Bataan Death March.

Lester Tenney's tank was knocked out during one of the many forgotten battles on the Bataan Peninsula. Captured by soldiers of the Rising Sun, he would survive the infamous Death March, suffer a deep wound from a Japanese officer's Samurai sword, cheat the Grim Reaper at Camp O'Donnell, and endure 30 days aboard a fetid 'hell ship' en route to a slave labor camp and captivity in Japan. Tenny faced nearly three years of pitiless labor in a coal mine owned by the Japanese company Mitsui. He recalled his injuries in captivity, "My nose was broken twice, my shoulder broken, my head split wide open, and my leg was broken." Slated for execution after being caught illicitly trading with Japanese guards, Tenney talked his way out of his own death.

The Battan Death March

Dr. Lester Tenney was guest speaker at the June, 2015, meeting of the Atlanta WWII Round Table. A remarkable speaker, no one attending was disappointed. The dining room was packed with members and guests, plus a heavy media representation. I have taken the liberty to use excerpts from his speech, his book "My Hitch in Hell" and from far-reaching research so Dr. Tenney can relate his experiences in his own words. Negligible revision was needed for clarity. One of only three American Bataan Death March survivors still alive, this is his amended story.

"Four straight months of combat. I was in the third of five

tanks sent in a straight line down this narrow road. The Japanese destroyed the tanks one by one, thus began my years of hell in captivity.

"For me, the Death March lasted 12 days, covering 68 miles. We had no food and no clean water. Men drank the filthy sludge out of buffalo wallows. I saw disembowelments, rapes, throats slit for no reason, soldiers beaten with rifle butts when they fell from exhaustion. Men were bayoneted, others beheaded. I lost a lot of friends, watched them bury a man alive. If you stopped, you died.

"An officer on horseback slashed my back with his sword but I couldn't stop, I had to keep going. I kept making goals for myself, the next knoll, the next herd of caribous or water buffalo, that next tree up ahead. At the town of San Fernando, they loaded us onto boxcars with no ventilation, no latrine, the heat intolerable, packed in like sardines. Men died standing up.

"After an agonizing four hours aboard the boxcars, we still had to stagger another six miles to Camp O'Donnell. After we arrived at the camp, my fellow countrymen stitched up the sword wound and treated the injury as best they could. It still became infected. I was bedridden battling the infection plus dysentery and malaria. I thought I'd die, but within days I was okay.

"The human suffering at Camp O'Donnell was bad enough but it was nowhere near the cruelty of a 'hell ship'. They herded us into the cargo holds of these vessels for the voyage to Japan. The conditions were wretched, overcrowded, no latrines, more men died standing up, the stench unbelievable. Dead men went unreported because we needed their rice and water rations.

"In Japan, they sent me to Fukuoka Camp 17 in Omuta, about 30 miles east of Nagasaki. The camp was eventually home to 17,000 Australian, American, and European POWs. We worked an old coal mine that had been shut down for safety reasons, but safe working conditions didn't apply to POWs. We were expendable animals. Routine beatings and torture were expected. One kid was tortured so badly he lost both legs. Guards beat you with boards, pickaxes, or shovels if they thought you weren't working hard enough.

"We labored in 12 hours shifts in the mine, morale plummeted; death was a relief. Somehow, I found the energy and bravado to ask the camp commander for permission to put on a musical show for the guys. I told him it would help morale and boost their production. He liked the idea. We called the production the "Ziegfeld Follies of 1944.

"The guards provided us with different colored beer bottles, green and red and blue, plus gave us cardboard. We busted up the bottles to use as 'jewels' and made paste from rolled rice. We glued the 'jewels' onto the cardboard to create headdresses, sewed a few brassieres, panties, and gowns, then found eight soldiers pretty enough to pass as women. Shoot, they looked good. They weren't proud gay, like in today's society, the guys were just…well, gay as heck, happy.

"The guys loved the show, hooped and hollered as if the fancy dressed G.I.s were the real thing. The camp commander enjoyed the show so much he allowed me to write home for the first time. It took over two years for the letter to arrive."

"I got caught bartering with the guards and was lined up for execution with five other guys. The commander came down the line, pointing at the men to be executed, 'You bayonet, you shoot,' but I was given the honor to be beheaded. Something told me, perhaps my gut instinct, that the commander didn't want to kill us, that perhaps he was having second thoughts. I started praising him for his discernable devotion to God and Country and I gave him compliments and respect. It worked. I talked my way right out of that execution and haven't stopped talking since.

"Three years of brutal day to day living. Cigarettes became the rate of exchange. We even formed a bankruptcy court so POWs could appeal their cigarette debts. Then one day a guard told us an enormous bomb had destroyed Hiroshima. We had no idea what happened, but a couple days later we saw a huge cloud boiling straight up with a mushroom top. It was 30 miles away at Nagasaki. We were speechless, yet knew the end was near.

"We speculated that four things would take place if the war ended: We'd get all the rice we desired, they would give us our Red Cross food boxes, we would not have to work, and we would no longer have to salute the guards. All that happened, plus one of the guards bowed to me. HE BOWED TO ME! People don't understand what that meant to us. I can still remember the camp commander's last words, 'America and Japan are now friends,' then left the camp with his soldiers.

"A short time later, this giant airplane flew over, circled around, then at very low altitude opened the bomb bay and parachuted containers full of food, water, and clothing into the camp. The plane was huge, a B-29, we'd never seen one. The next day we shared our rations with starving Japanese outside the gate. I didn't have hatred in my heart; the more I associated with the Japanese, the less hatred I had. The Bataan Death marchers are nearly all gone now, I think three of us are left. I hope and pray my fellow Americans never forget what freedom really means."

After the war, Tenney returned to school and earned a PhD in finance. He taught at Arizona State and San Diego State. Dr. Lester Tenney is 96 years old, his wife Betty is 97. As he stated during his presentation, "Betty likes younger men."

Dr. Tenney slipped gently into the Good Night on February 24, 2017. One of America's greatest examples of The Greatest Generation is no longer with us, nor his lectures, his wit, his compassion, and his incredible ability to educate our younger generations on the true meaning of freedom. He believed in individual responsibility and related how freedom was gained, protected, and saved at such a terrible cost. He will be sorely missed doesn't begin to describe our country's loss.

SC-1031: A LITTLE SHIP WITH A BIG HEART

In 1828, the two Helms brothers received a land grant for a homestead in Henry County, GA. They packed their belongings, hitched up an old blind mule, loaded the kids into a wagon (both had lost their wives) and began the arduous journey from the Carolinas to their new habitat. Once settled, they built a log cabin and worked the acreage. In 1870, their farm was assimilated into a newfound county called Rockdale. Helms descendants have settled Rockdale County ever since.

Albeit one descendant, Jack Wilson Helms, and his future wife, Dorothy Virginia Smith; were born in Atlanta. Dorothy recalled, "Jack was 11 years old when he moved away after his mom died, but in 1940 he moved back right down the street from me. We both had sparks in our eyes. We wed on April 10, 1941."

Emotionally recalling Pearl Harbor, Dorothy said, "We knew it meant war, but our biggest concern was Jack's older brother, Brian." Brian's ship, the USS Vestal, was moored alongside the doomed battleship USS Arizona on Dec 7, 1941. When the Arizona blew apart, the explosion cleared the deck on the Vestal. The Arizona took 1,177 sailors and Marines down with her, but miraculously only seven sailors aboard the Vestal were lost.

Dorothy continued, "Brian finally called his father. He was safe. He'd spent the night on shore."

Dorothy's husband, Jack, joined the Navy on May 18, 1942. He trained in Miami as a Navy Quartermaster and quickly learned the tricks of the trade in ship navigation and the maintenance of nautical charts and maps. In his spare time, Jack jumped into the ring as a Navy boxer. Eventually transferred to the Atlanta Naval Air Station, currently Peachtree-DeKalb Airport for short-term duty, he waited for the completion of his assigned ship: Subchaser 1031, or SC-1031.

Jack Helms, U.S. Navy

Nicknamed the "splinter fleet" due to their wooden hulls, the tiny sub-chasers had a crew of about 25. Loaded with detection and ranging equipment, the standard armament consisted of dual-purpose 40mm guns, 20mm guns, and depth charges.

Crew of SC1031

Dorothy recalled, "We enjoyed the Atlanta Naval Air Station but we realized he'd soon be going to war." Commissioned in New Orleans on Feb 4, 1943, the SC-1031 first cruised through the Panama Canal then voyaged up the West Coast to Alaska for the American effort to retake Japanese-occupied Attu and Kiska Islands in the Aleutians.

"Jack almost froze to death up there in the Aleutians," Dorothy said. "One of his shipmates became so depressed from the cold he attempted to jump overboard, but Jack jerked the man back, saving his life. Later on, Jack was nearly washed overboard in rough seas, but another sailor saved him. I guess their Guardian Angels were working overtime."

Allied soldiers assaulted the western most Aleutian Island,

Attu, on May 11, 1943. The harsh terrain and fanatical Japanese took a heavy toll on American soldiers. In desperation, the Japanese executed a suicidal 'banzai' charge, one of the largest of the war. The fighting was personal, much of it hand-to-hand. After the melee, 580 allied soldiers lay dead and hundreds more were wounded from two deadly adversaries: the Japanese and sub-freezing temperatures. The banzai attack annihilated the remaining Japanese on Attu: 2,351 dead with hundreds missing. Only 28 enemy soldiers were captured.

Dorothy confirmed the gruesome cost of victory: "When Jack went ashore he saw hundreds of bodies stacked up like cords of wood. That really upset him." Ships, large and small, suffered from the unforgiving weather. Footing became hazardous; men succumbed to frostbite; guns froze; equipment failed.

August 15, 1943: A force of 35,000 allied soldiers land on Kiska Island to engage a phantom army. Under the cover of thick fog, the Japanese had managed to evacuate their soldiers on July 28 and slipped away undetected. Nevertheless, the allies suffered in excess of 300 casualties from booby traps, disease, frostbite, and 'friendly fire' incidents.

Jack Helms and SC-1031 left the cold hell of Alaska and sailed south into the hot hell of the Pacific. Jack had the opportunity for a social visit with his brother, Brian, during a layover at Pearl Harbor. Dorothy recalled, "Brian was Jack's senior by 12 years, so they hadn't seen each other in a decade. Jack went aboard Brian's new ship, an escort carrier, to reunite with him. Well, they met, but Brian had no idea who Jack was and asked to see his

I.D. Jack showed his I.D. then they had one heck of a good time together."

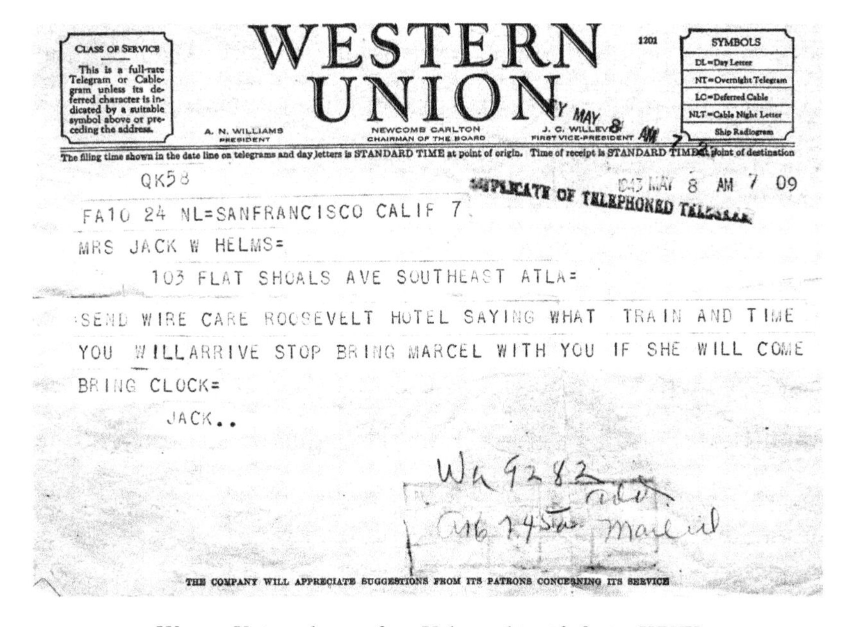
WESTERN UNION

CLASS OF SERVICE
This is a full-rate Telegram or Cablegram unless its deferred character is indicated by a suitable symbol above or preceding the address.

SYMBOLS
DL=Day Letter
NT=Overnight Telegram
LC=Deferred Cable
NLT=Cable Night Letter
Ship Radiogram

A. N. WILLIAMS PRESIDENT — NEWCOMB CARLTON CHAIRMAN OF THE BOARD — J. C. WILLEVER FIRST VICE-PRESIDENT

The filing time shown in the date line on telegrams and day letters is STANDARD TIME at point of origin. Time of receipt is STANDARD TIME at point of destination

QK58 1943 MAY 8 AM 7 09

DUPLICATE OF TELEPHONED TELEGRAM

FA10 24 NL=SANFRANCISCO CALIF 7

MRS JACK W HELMS=

103 FLAT SHOALS AVE SOUTHEAST ATLA=

SEND WIRE CARE ROOSEVELT HOTEL SAYING WHAT TRAIN AND TIME YOU WILLARRIVE STOP BRING MARCEL WITH YOU IF SHE WILL COME BRING CLOCK=

JACK..

THE COMPANY WILL APPRECIATE SUGGESTIONS FROM ITS PATRONS CONCERNING ITS SERVICE

Western Union telegram from Helms to his wife during WWII

Dorothy stated during the interview that Brian's escort carrier was sunk during the Battle of Leyte Gulf. Two escort carriers were deep-sixed during the battle, the *Gambier Bay* and the *St. Lo.* Brian, however, was rescued and survived the war.

SC-1031 and her Quartermaster Jack Helms also sailed into a hotbed of activity. The sailors aboard the little vessel played a part in the Gilbert and Marshall Islands campaigns. The Gilbert Islands were the first targets with bloody battles on Tarawa and Makin. The Marshalls were next. Fierce combat occurred on land, air, and sea around atolls or islands with names like Mili, Kwajalein,

Majuro, Enewetak, and Utirik. Pacific Islands in the middle of nowhere, fought for and paid for with the lives of young men who couldn't even pronounce the strange-sounding names.

SC-1031's war record is lost to history. Her assignments placed the vessel in Harm's Way throughout the war, yet like so many other ships and airplanes and missing warriors, too much has been misplaced or lost concerning their service and may never be recovered.

Helms Brothers Construction Company was Jack's civilian port-of-call. He started a family and did what men of the Greatest Generation were expected to do, produce, build, and stay loyal to his faith and to the nation for which he fought. Jack and Dorothy bought a farm in his ancestor's old stomping grounds: Rockdale County, GA. Dorothy said of Jack's farming techniques, "Shoot, he wanted to raise Black Angus cows but I surely don't know why. He treated those cows like they were his pets. They were the most spoiled cows in Rockdale County."

Extensive research failed to discover a photo of SC-1031, even though some likely exist. Maybe one or two occupy an old shoe box in a spare closet or yellow-away at the bottom of a dusty trunk in a dustier attic. Cherished heirlooms have a bad habit of losing their appeal to younger generations, and lost with the apathy is another story left untold. Near the end of WWII, the little ship with the big heart was donated to our ally, the Soviet Union. Her fate is unknown.

SC-1031's stalwart Quartermaster, Jack Helms, reported for his final inspection on February 6, 1998.

Only known photo of SC-1031

A LITTLE SHIP THAT COULD

Habitually identified as the Splinter Fleet, the tiny 110-foot wood-hulled Sub Chasers of WWII held the title as the smallest commissioned ship in the US Navy. A Sub Chaser cruised at 12 knots with flank speed no more than 20 knots. The more popular PT-Boats of John F. Kennedy's PT-109 renown were only 80 feet long but often hit 40 knots, plus PT-Boats were commissioned collectively in squadrons, not individually.

Comparable to a roller coaster ride even in placid seas, seasoned sailors aboard Sub Chasers were often seasick, especially after an extended shore leave. The crew of three officers and 24 enlisted men utilized a 3" deck gun and twin .50 caliber machine guns for their protection. Later in the war, the 3" deck gun was replaced with a Bofors 40mm gun and two 20mm guns replaced the .50 caliber machine guns. To combat submarines, the crew dropped depth charges or used a forward firing rocket contraption called a Mousetrap Rack.

On April 16, 2013, the newspapers published my article on the Helms brothers, Brian and Jack, both Navy veterans of WWII. Brian was aboard the USS *Vestal* during the Japanese attack on Pearl Harbor and dodged the Grim Reaper a second time as a crew member on one of the two escort carriers sunk during the Battle of Leyte Gulf. Brian's younger brother, Jack, served in the Pacific aboard a Sub Chaser, SC-1031. Somewhere in a grubby attic in a grubbier storage box perhaps the war records of SC-1031 may exist, or maybe her history is lost forever. We know for a fact that Jack Helms fought in the freezing seas around the Aleutian Islands and in the tropical humidity of the Marshall and Gilbert Islands, yet little is known regarding the events. A photo of SC-1031 was not to be found.

April of 2015: I received an email out of the blue from a Mister Jeffrey Fisk, a lawyer in Chicago. His father, Robert (Bob) Fisk, Jr. served aboard SC-1031 during WWII. He discovered my article on SC-1031 and the Helms Brothers by Googling "News" in an ongoing research project to find all the information available on his father's ship and service. Twenty years elapsed before Jeff discovered a photo of SC-1031 and, by a stroke of luck, googled my story.

Therefore, this is the story of two sailors, Jeff Fisk's father, Bob Fisk, Jr., and his grandfather, Bob Fisk, Sr. The grandfather, Bob Fisk, Sr. lied concerning his age and joined the US Army when he was 15 years old. By 1916, he had achieved the rank of Master Sergeant and served as a bugler under General John "Black Jack" Pershing during the American raid into Mexico to search for the revolutionary and bandit Pancho Villa.

In El Paso, Fisk, Sr. saddled-up a horse fresh from the prairie to participate in a parade. The panicky horse buckled then fell on the cobblestones in front of General Pershing's reviewing stand. Fisk, Sr. calmed the horse then finished the parade. Called back to the reviewing stand, Fisk, Sr. endured harsh words from his Commanding Officer, utterances not to be repeated in a family newspaper. After the Commanding Officer ran out of breath and offensive language, Fisk, Sr. glanced up at General Pershing and stated, "Those are the nicest words he's ever spoken to me."

The elder Fisk decided to leave the Army and join the Navy. He served on a submarine tender USS *Rainbow* off the Atlantic Coast during WWI and later sailed on a supply ship to China only to be commandeered by the Japanese after their invasion of China in 1937. Fisk was stranded in China for over a year. When finally released, he returned to the States to join the Merchant Marines in WWII, working on the oceangoing tug USS *Boone Island.*

Now another Fisk desired to 'see the world'. Bob Fisk, Jr. was itching to join the fight. While still in high school, he asked his grandparents to sign the papers since he was underage. They refused; education came first. In 1943, high school graduate Bob

Fisk, Jr. joined the Navy, took his basic training on the Great Lakes, arrived in San Francisco in December '43 then boarded a troop ship for Pearl Harbor. One of the smallest ships in the Navy was waiting for him, Sub Chaser SC-1031.

Bob Fisk, Jr. - U.S. Navy

BOB FISK, JR'S STORY
AS TOLD BY HIS SON, JEFFERY FISK

"I remember my Dad telling me about December 7, 1941 and a ten block long line the next day at the recruiting station. Dad wanted to fight, but my great-grandparents were unwavering about him finishing high school. Our entire family contributed; I even had one Aunt who worked on Dauntless Dive bombers in Corpus Christi.

"Dad said the skipper of SC-1031 was gung-ho and wanted back in the action. Apparently the ship had seen plenty of action, but was pulled back to Pearl for reasons unknown. Well, the gung-ho skipper did get back in action but not aboard SC-1031. A fresh skipper took over the helm. The officer was good friends with an Admiral and talked his way into milk run patrol duties around the Hawaiian Islands.

"So, that's what Dad and his shipmates did, most of the time. The ship was sent on missions and supply runs to the Marshall Islands and many more, but the experience was very sobering. The Hawaiian Islands were so lush and pristine, but Dad said the Marshalls and other islands had been devastated, no palm trees, little vegetation, the islands looked like the surface of the moon. War isn't pretty.

"He recalled seeing other combat ships limp into Pearl, some listing so badly they were hardly afloat. Dad said you could hear all the activity on the ships before they made port, the welding, the hammering, the cutting of steel, then the ships went into dry dock.

Around 90% of the welders were Nikkei Amerikajin, which means Japanese American. There were no internment camps in Hawaii. Dad said their welds were so smooth it looked like they had been put on with a butter knife. The damaged vessels were repaired and put back to sea, usually within three to four days.

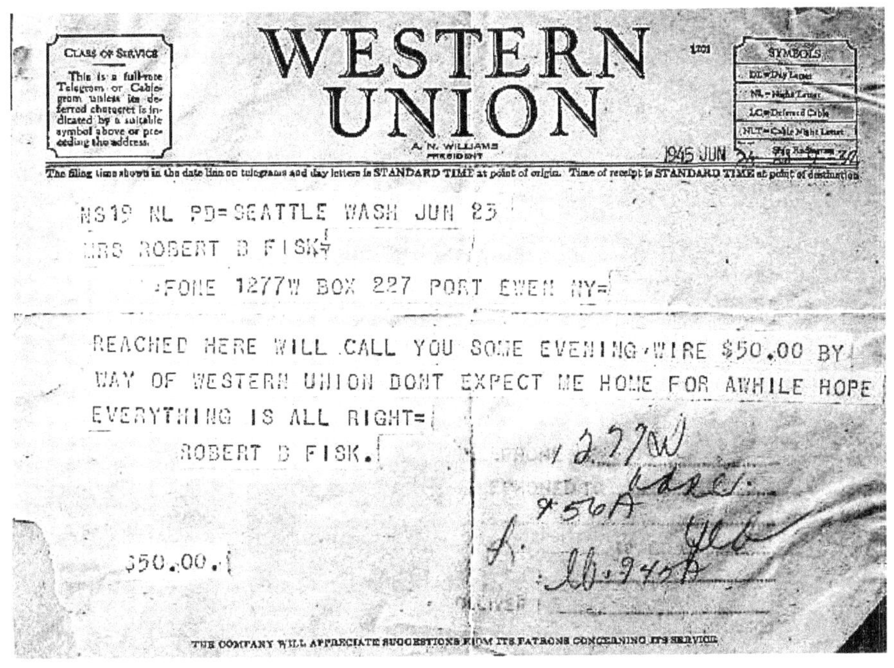

WESTERN UNION

NS19 NL PD=SEATTLE WASH JUN 23

MRS ROBERT D FISK

=FONE 1277W BOX 227 PORT EWEN NY=

REACHED HERE WILL CALL YOU SOME EVENING WIRE $50.00 BY WAY OF WESTERN UNION DONT EXPECT ME HOME FOR AWHILE HOPE EVERYTHING IS ALL RIGHT=

ROBERT D FISK.

$50.00.

Western Union telegram from Fish to his wife during WWII

"Dad didn't see combat. He said his service on the SC-1031 was a precursor for McHale's Navy. They had a dog on the ship, the Captain had a Harley Davidson Motorcycle, and my Dad owned an Indian Motorcycle. They carried their bikes around on the back of the ship from island to island so they could explore or whatever.

Someone finally ordered them to remove the bikes from what was alleged to be a combat vessel.

"Dad always said that he was lucky and counted his blessings. He helped with the transition of SC-1031 to the Russians and also assisted in training the Soviet sailors for the upcoming invasion of Japan. After the transition, he was en route home on a troop ship when the war ended.

"Back in San Francisco, Dad truly believed the Naval Department had gone bonkers. They put him on a train from San Francisco to New York to pick up his new orders: to serve on a yard oiler, *YO105*, in San Francisco. So, my father caught another train from New York back to San Francisco. Ya gotta love the military."

Bob Fisk, Jr returned to civilian life in 1946, worked at a boat yard for 10 years until he found employment at International Business Machines as a senior electronics technician. He worked in the SEDAB Department (Special Engineering Design and Build) and was instrumental in the development of IBM's first System One Business Computer. He retired after 32 years with IBM.

I connected Enid Hanson, daughter of SC-1031 crewmember Jack Helms, and Jeffrey Fisk, son of SC-1031 crewmember Bob Fisk, Jr., with each other's email addresses and phone numbers. They met in Atlanta within a year to discuss their fathers and a tiny wooden-hulled Sub Chaser. The 'kids' thanked me for making this connection happen but I was just a small cog in a big wheel. The tiny Sub Chaser SC-1031 was the real matchmaker, the Little Ship That Could. Her fate with the Russian Navy is still unknown.

UNQUESTIONABLY PATRIOTIC

> *"None of us here in Washington knows all or even half of the answers. You people out there in the fifty States had better understand that. "If you love your country, don't depend on handouts from Washington for your information. If you cherish your freedom, don't leave it all up to BIG GOVERNMENT."*
>
> —Senator Barry Goldwater, 1962

From the Random House College Dictionary:

Capitalism: An economic system in which investment in and ownership of the means of production, the distribution, and exchange of wealth is maintained and made chiefly by private individuals and corporations.

Socialism: A theory or system of social organization advocating the ownership and control of industry, capital, land, etc. by the community as a whole. In Marxist theory: the stage following Capitalism in the transition of a society to Communism.

Utopian Socialism: Founded upon or involving imaginary political or social perfection. Given to dreams or schemes of such perfection.

Communism:A theory or system of social organization based

on the holding of all property in common, the actual ownership being ascribed to the community as a whole, or, to the state. A system in which all social and economic activity is controlled by a totalitarian state dominated by a single self-perpetuating political party.

*****

I have talked to thousands, interviewed hundreds, listened to dozens of lectures or presentations made by American veterans, yet not one ever made the claim they fought to defend socialism or any other ill-conceived form of government other than capitalism and the United States of America. They also fought for Old Glory, so she could continue to fly with pride and dignity above our great nation. And I have never heard a veteran claim that he or she fought for the twisted concept of political correctness so misguided individuals could whine about how offended they are by the Stars and Stripes and demand that she be hauled down.

What other country will haul down their national flag because of seditionist rabble-rousing characters claiming to be 'offended' by the national colors of whatever country they may be living in? Com'on, name one. How about Russia? France? Switzerland? Kenya? Panama? New Zealand? What's the matter, pussycat got your tongue?

Another befuddlement veterans have trouble coping with is the money-wasting ridiculous debate in Congress over 'border security'. The vast majority of veterans cannot fathom a government that would be eccentric enough to even conduct such a debate. Again, what other sovereign country allows illegal open-boarder entry into their territory after which the same host country feeds, clothes, employs, provides medical care, awards them a driver's license, plus allows municipalities to ignore with condescension long-standing federal laws by offering illegal aliens hiding places called sanctuary cities? Too, what other supposedly logical country will bring up a vote to their governing body of elected representatives in a clandestine attempt to have the taxpayers they supposedly represent pay the tab for granting funds to children of illegal aliens for a college education that the same disrespected taxpayers can ill-afford for their own children?

Go ahead, sneak into Spain then demand in the streets if necessary that the Spanish government support you, educate and feed your children, grant you spending money, and see to it that they haul down their national colors due to the fact that you're offended. When Spain kicks you out of their country, move on to, oh, let's say, Palestine. Sneak in during one of the many rocket attacks on Israel and request the same. Okay, now the officials in Palestine have booted your ass out from their turf because they don't intend to waste money nor time keeping you in a prison or insane asylum. I guess you need to move on to, well, let's stay around the Mediterranean environs, so, how about Italy? Wonderful, and while you're in Italy hike up your uncouth demands to include the Pope. March to silence his beliefs in defense of your agnostic beliefs. I mean, why not? The world *is* all about *you,* isn't it?

What makes an individual so egotistical as to think a sovereign country should, could, or would cater to their own personal whims? Refer to the quote below.

"There is nothing more frightening than ignorance in action."
—Goethe

Frustration with politicians, non-producers, weak-minded intellectuals lacking the common sense to pour piss out of a boot, better-than-thou bums, and protestors without value has reached the breaking point. America has gone on the defensive to defend its very existence. Anxiety reigns. Liberals have replaced logic, communists have replaced conservatism, socialism has replaced sovereignty, and political correctness has replaced patriotic countrymen. Educated law-abiding citizens carrying the American flag are called Radicals. Peaceful demonstrators that are worried sick for their country are referred to as Astroturf. And the icing on the political cake: the comedians running our Department of Homeland Defense recently sent out a warning to law enforcement agencies throughout America to be on their guard for potential terrorists: our returning veterans from Iraq and Afghanistan.

It's laughable; it's ludicrous, and lame-brained. Thank the Lord, sensible Americans understand what is at jeopardy: Our freedoms, fought for and won by veterans.

WWII combat veteran Grady Mullins recognizes the threat. His dismay is echoed by WWII veterans of all races, colors, and creeds by one simple fact: Freedom is not free. It never has been; it never will be. Freedom requires blood and lives. It demands bravery of nineteen-year-old B-17 pilots; it takes the resolve to live and fight rather than succumb to a freezing death on a mountain peak in Korea, and it withstands a two-and-a-half month siege on a remote plateau at an old French airstrip in Vietnam called Khe Sanh. Freedom takes courage; unyielding values, and individuals willing to scream, "By God, enough is enough!" Or as Grady Mullins would demand, "Stand up and be counted!"

Grady Mullins - WWII Combat Veteran

Unquestionably patriotic, Grady spends his retirement fighting the good fight. His newspaper articles pull no punches on the good of America. His emails plead for local political involvement by an apathetic citizenry. His words of warning sting the faces of Americans still asleep of the danger. "I don't understand," Grady said. "People don't even seem to care about what's going on. I think they'd rather play with their cellphones than defend our independence." A warrior from The Greatest Generation, his eyes reflect the mirrors of the war, both foreign and domestic.

> *"Look at an infantryman's eyes and you can tell how much war he has seen."*
>
> —Bill Mauldin

My father and Grady Mullins have some things in common. Dad was born and grew up in Dunmore, Pennsylvania, the son of Italian immigrants. Pop was the youngest of twelve siblings and served with the Army Air Corp in India and Burma during World War Two. Grady Mullins was born and grew up in the rural farm community of Jasper, Georgia, also the youngest of twelve siblings. Back in those days, large families ensured economic survival and the potential to go forth and multiply the family genes. Political correctness meant folks voted on election day, a cell phone was the only communication device at the local jail, I-pod was a slurred pronunciation for stinky stuff in the cow pasture, a stiff summer breeze air-conditioned your home, 'pot belly' described the heating apparatus in your living room and not a Budweiser Belly, and 'famous gay couples' meant good-humored slapstick comedians like Ozzie and Harriet Nelson or George and Gracie Burns. Things were, well, different.

Grady explained, "We were dirt poor farmers and what we grew on the farm, by golly, we ate. All of us youngins' had chores and I wish I had a dollar for every cow I milked. Young people now don't know how bad things were back then. You had to grow up fast, real fast, but in my mind, I didn't become a real human being until June 28, 1942."

Erroneously believing I had a handle on his last comment, I suggested, "Let me guess, Grady, that's when you joined the military, right?"

"Nope," he replied. "That's when I turned sixteen and got my social security card and my Georgia driver's licenses." Some things, well, aren't different.

On December 7, 1941, the Mullins' family radio informed them a place called Pearl Harbor had been bombed by the Japanese. Most of the people around Jasper didn't know where or what Pearl Harbor was. But they soon found out it meant war. Of five Mullins brothers, four went to war.

"It's hard for folks today to realize what effect that had on families," Grady said. "Your father, your brothers, your cousins, your sons, it seemed everybody you knew put on the uniform. Even the wives and mothers saved bacon grease to be used in explosives. We were all in this thing together."

Too young to enlist after Pearl Harbor, Grady took a job in Atlanta at an iron works plant building huge steel double-doors for LST's (landing ship tanks). But on his eighteenth birthday, Grady followed his brothers into a world war. At 6'5" in height, he was too tall for Army aviation and ended up as a ground-pounder. In September of 1944, he entered the seventeen week training course for combat infantry but a German offensive known as The Battle of the Bulge that December cut short Grady's training. He received a five day leave to visit his parents. On the fifth day, they escorted their youngest son to the train station in Atlanta and watched him board a Pullman car.

"I was the last son to go," Grady said. "I still remember the bleak look on my mother's face and it was the first time I saw tears in my father's eyes."

Eventually assigned to the 66th Infantry "Black Panther" Division, Grady crossed the Pond in comfort on the *Ile De France*, at that time the third largest luxury liner in the world. Another vessel, the converted Belgian passenger ship *SS Leopoldville*, transported another 2500 men of the Black Panthers across the English Channel. Five miles short of the port at Cherbourg, the *SS Leopoldville* was hit and sunk by a torpedo from a German submarine. Fourteen officers and seven hundred forty-eight enlisted men lost their lives.

Grady Mullins - US Army, WWII

After landing in France, the 66th Black Panthers were sent into a blocking position to suppress and/or eliminate over 100,000 cornered but battle-hardened German soldiers in the Lorient and St. Nazaire pockets. The battle was a brutal but little-known mêlée far behind the Allied advance across the Rhine.

1944-1945

Grady Mullins - The Terror of Europe

Grady recalled, "The Germans had 88mm artillery mounted on trucks. They'd back up the trucks to the top of a hill and cut loose on us. In one fight, a piece of shrapnel nicked my leg and embedded in a tree. I cut the piece of shrapnel out of the tree and kept it. I still have it to this day." But no Purple Heart for Grady; he returned to combat after the wound was patched with a Band-Aid.

In the snow and frigid cold, the Black Panthers continued their aggressive tactics. Artillery duels were common and deadly. During February of 1945, the 66th fired an average of 1,140 rounds per day. In March, an average of 2,000 daily rounds were fired on German positions. In April, the Black Panthers lobbed 66,000 shells at the beleaguered enemy.

A captured German had a copy of the movie schedule for the largest theater in Lorient. Operating on the schedule, the next night the 66th Field Artillery destroyed the theater. The Germans showed their films in concrete bunkers after that night.

"I fired my M-1 all the time but don't know if I ever hit anything," Grady said with a smile. "We'd plant booby traps and mines, then the Germans would dig them up at night and lay them out in a neat little line just to aggravate us the next morning. Grown men were dying out there, yet grown men were still playing jokes on each other."

After Germany's surrender, the 66th traveled 700 miles to occupy 2400 square miles of Reich territory. The Black Panthers had hardly settled down when ordered back to Southern France to help billet, feed, and process troops redeployed to the Pacific through the port of Marseille.

"That was sort of a lucky break," Grady said. "The army had a point system and I had enough so-called "points" to keep me out of the Pacific. Plus, due to my processing experience, the Army offered me $350.00 to re-enlist for another year to help process out the soldiers coming home from the war. That was a lot of money back then." Grady was sent to Camp Campbell, Kentucky for his last year of military service to process out our returning veterans. He is the recipient of the Bronze Star and two campaign ribbons, Grady Mullins was discharged in December of 1946.

He worked in his brother's road construction business after the war. "Let me tell you something about asphalt," Grady said. "It's hot and it's heavy." Striking out on his own in 1955, he went to work for a commercial airline company. "The company got its start crop-dusting in Louisiana," he stated. "Today people know it as Delta Airlines."

Grady retired from Delta in 1989. A poor farmer; an iron worker; a combat veteran; a construction man; a retiree from a great American corporation; another great example of the Greatest Generation.

A footnote: Grady discovered a hidden talent in his Golden Years: he's an absolute terror on the felt of a poker table. It seems you just can't keep The Greatest Generation from a new challenge.

THE KOREAN WAR

"I fired him because he wouldn't respect the authority of the President. That's the answer to that. I didn't fire him because he was a dumb son-of-a-bitch, although he was, but that's not against the law for generals. If it was, half to three-quarters of them would be in jail."

—Harry S. Truman, on his dismissal of
General Douglas MacArthur

THE FORGOTTEN SOLDIERS

"The object of war is to survive it."

—John Irving

Historically the Korean War has been known as "The Forgotten War", which, in my opinion, is a disrespectful suggestion that the misery and sacrifices of the soldiers who fought in that atrocious war are somehow ignored. As a free-lance writer and Vietnam veteran, I'm offended that such a second-rate allusion has been constantly repeated. For the soldiers who survived Korea, the vivid nightmares and memories will never be 'forgotten'.

It's feasible, I reckon, that the enduring suggestion of 'forgotten' is a sign that a nation demanding an 'unconditional surrender' in WWII was unable to cope with the concept of a 'limited war' or a 'police action' with one goal: 'Containment' of Communism. More imaginably, the suggestion may indicate exactly what it proposes: Forgetting the Korean War and chalking it up as a no win/ no lose struggle that will not soon be repeated.

The Korean War was the facilitator of change from 'There is no substitute for victory' to a flaccid foreign policy based on power restraint and the irresponsible notion of having 'empathy' for our enemies. There ain't no 'nice guy' in war. War is a dirty, nauseating enterprise and should always be the last resort of political failure. But when the military is given the order to 'go,' then let them 'go' do what they've been trained to do. Cut the head off the snake, let the body wither and rot, then bring our boys home.

COACH HAMBRICK

Georgia educator Charles Ray Hambrick coached basketball and taught Physical Education at Briarcliff High School, served as an administrator in the Tift County school system, and worked as an assistant principal of Rockdale County High School. The students and faculty had no idea that Hambrick was a decorated combat veteran of the Korean War.

After graduating from high school in 1950, Hambrick's football and basketball talent earned him a scholarship to West Georgia

College. Albeit, North Korean dictator Kim Il Sung's attempt to overpower South Korea put many a young man's education on hold. On June 25, 1950, North Korean forces stormed across the 38th Parallel dividing North and South Korea. America, via the United Nations, found itself in another war.

Hambrick in Korea

Hambrick recalled, "My friend, Walter Mitchell, and I decided to join the Army in 1951, but lost contact with each other after basic training at Fort Jackson, South Carolina." Gung-ho, athletic, and undaunted by a challenge, Hambrick volunteered for airborne training at Fort Benning, Georgia. He continued, "The first time I jumped out of a plane I thought '*Ray, I believe you've made a big mistake*'. I felt a lot better after the parachute opened." After earning 'top graduate' status at jump school, the Army asked Hambrick to stay at Fort Benning as an instructor. "I turned them down," he said. "That was another big mistake." Within two days, Hambrick was en route to Korea.

FOOTBALL FIELDS TO KILLING FIELDS

Seasick for 27 days en route via a WWII Liberty Ship to Sasebo, Japan, Hambrick and his buddies boarded a flat bottom barge for transport to Korea. "That was another up-chucking voyage," he admitted. The pancake-flat barge docked at Pusan in December of 1952, smack dab in the middle of a cold winter. From there, Hambrick boarded a troop train for the capital city of Seoul. His memories of the South Korean capitol remain graphic. "Seoul was flat, utterly destroyed, and the weather was subzero. I remember how cold it was, yet the starving children kept running up to our train for handouts with no clothes or shoes on."

Hambrick's first assignment put him in harm's way almost immediately. He recalled, "I was sent into combat near the 38th parallel that divides North Korea from South Korea. I trudged up a hill in snow up to my butt cheeks. When I got to the top of the hill, a sergeant barked at me to find a hole to get a little sleep. Well, the hole I slept in turned out to be the hole they used for an ammo dump. I quickly found another hole." His weapon of choice was the M-1 Carbine. His reasoning made sense: "The M-1 Garand only held an 8-round clip, but the carbine could hold a 30-round banana clip with another clip taped to it for a quick swap out. We used a lot of ammo in Korea."

Hambrick, right, in Korea

Day in, day out, nothing changed in Korea, not even the frigid weather: "We just stayed alive, that's about it," he said. "I'm certainly no hero; I'm a survivor." The brutality of the war is still embedded in Hambrick's nightmares. "When the Chinese captured an American, they staked him out in the freezing cold, naked as a Jaybird, and we couldn't rescue these guys. That's tough to forget." His leadership stopped an enemy attack as reported in a featured article in his hometown paper on February 5, 1953: "*A platoon sergeant credited Sgt. Hambrick with stopping the enemy attack. Hambrick had a crew on target within 30 seconds and kept the Reds penned down.*"

Hambrick, left, in Korea

After nearly six months of ruthless combat, Hambrick received orders to report to headquarters command. "They offered me a deal," he recalled. "Go back to jump school but extend your enlistment. Shoot, I refused, and they sent me right back to the front." Grinning, he admitted, "Well, I guess that was another mistake." Two days later, headquarters offered him a better deal. "They want me to attend helicopter school, which offered a warrant officer's commission. Since my tour was almost up, I refused. I view that as another mistake because I found myself right back on the front, again."

Without any meaningful training, Hambrick and other soldiers were suddenly taken off the front and assigned to Koje-Do Island to guard Chinese prisoners. But the killing continued. Hambrick said, "During a Chinese holiday, about 3,000 POWs rioted. A U.S. Army Captain tried three times to quell the riot, but when a Chinese leader refused to do as ordered on the third attempt, things really got out of hand. The Captain pulled his .45 pistol from his holster then blew the guy's head off. Man, all hell broke loose. I was manning a guard tower and had to cut loose on the Chinese when they charged my position." Pausing a moment, Hambrick stated, "Folks don't understand the lasting effect of taking a human life. I understand that it's kill or be killed, but it's still killing."

Hambrick's thoughts on not bathing for three months: "Never been so filthy in my life. Sometimes we would wash with fresh snow but at 20 degrees below zero it didn't accomplish very much." Finally, after three months, he was sent to the rear area for a hot shower and clean uniform. Describing the trip back to the front, "We rode back in an Army truck and ran into a blizzard. It was snowing like crazy. I put a towel over my head then put my helmet back on for a little protection. When we reached the front, the towel was frozen solid, like concrete. I almost pulled my ears off getting free of that towel."

On one occasion, Hambrick volunteered to rescue two soldiers ensnared in a mine field. "I reckon that was another mistake," he admitted. "I stepped on a land mine trying to get the men out. I heard the 'click' and figured that I was a goner. Thank the Lord, the second 'click' never happened." By luck or the Lord, Ray Hambrick was spared and two frightened soldiers were finally pulled to safety by the courageous sergeant from Georgia.

A year passed; Hambrick's war was finally over. He stated, "I was sent to Pusan to catch a homebound ship. The first person I ran into on the ship was my friend Walter Mitchell. He had made it, too. We partied like there was no tomorrow."

Back in the real world of steak and round-eyed girls, Hambrick trained raw recruits at Fort Lee, VA until orders came down for Rhein-Main, Germany. Perhaps rewarded for surviving the hell called Korea, Ray Hambrick spent his last year in the military doing what he enjoyed the most: playing football and basketball for the United States Army. His athletic talents helped win a conference championship.

His duty done, Hambrick returned to West Georgia for a couple of years on the G.I. Bill where he met his future wife, Jean, but a knee injury on the football field of dreams kept him from achieving his lifelong quest: playing 'between the hedges' at the University of Georgia. He remained on campus and earned a B.S. Degree at U.G.A. then acquired his Master's Degree in Education from Georgia Southern.

Charles Ray Hambrick is the real John Wayne, an unsung hero that did his duty, came home to educate, and tried to forget a battlefield called Korea. Now retired, he spends leisure time on local golf courses shanking or slicing his way through 18 holes or visits his favorite fishing hole to drown the contents of a live bait container. One activity he loathes: "I have over 20 acres on my property so I spend far too much time on my tractor." When asked if the surplus acreage yielded anything, he stated, "Yeah, taxes."

THE WAR IN VIETNAM

"We should declare war on North Vietnam. We could pave the whole place over by noon and be home for dinner."

—California Governor Ronald Reagan
quoted in Edmund G. Brown
Reagan: The Political Chameleon, 1976

DEEDS NOT WORDS

Words mean a lot to politicians, not to soldiers. Soldiers' lives depend on deeds, not words, not promises, and certainly not on crippling rules of engagement. Young lieutenants in Vietnam understood the weight of words, an erroneous command, a misunderstood order, telling your men to take the wrong jungle trail, calling in wrong coordinates. Bob Babcock knew the appropriate words to motivate his soldiers, but their deeds decided who lived or who died. Deeds not words best describes Bob Babcock the man, and his experiences in Vietnam. This is his story.

"I was born and raised in a railroad town of 2,000 folks called Heavener, Oklahoma, essentially a terminal for the Kansas City

Southern Railroad. My grandfather and my father, shoot, about everybody in town were railroaders. After high school, I attended what is now Pittsburg State University in Kansas where I took ROTC and was a distinguished military graduate. I applied for three areas of training in the Army, infantry as the first, and that's exactly what I got. It made sense, in 1965 infantry officers were in demand."

Babcock was sent to Fort Benning, Georgia for his Infantry Officer basic course. Of course he went airborne and jumped out of perfectly good aircraft. When asked why, he responded, "Well, I wanted to prove to my buddies and myself that I wasn't a coward, so I did it. I still remember my first jump, scared going out and excited that I was still alive when I landed. I made my jumps to qualify and never had a desire to do it again."

Fort Lewis, WA—Thanksgiving Day, 1965: "We were understrength, no doubt about it. I was in the 4th Infantry Division in a platoon of eight soldiers; I was authorized 43. We were assigned to Task Force Ace, a ready to go anywhere in the world type of unit that rotated between the 82nd, 101st, and 4th Infantry Divisions. Well, we had an alert and suddenly my platoon is up to 40 men, 32 of which I'd never seen before. We went aboard a C-97 Stratofreighter to, well, to somewhere. Then they called off the alert, except for us, we stayed on board. We circled around for a bit, landed at Gray Army Airfield at Ft. Lewis, then attacked the hill at the end of the airfield during rush hour. We'd been up since 0200 and now it is late afternoon, we're exhausted. The exercise ended so we went back to our quarters, cleaned our equipment, and then those new guys disappeared. We never saw them again."

Bob Babcock in Vietnam

"Then I'm taken off Task Force Ace and start getting new troops right out of basic. We were told to give them advanced individual training and get them ready. Ready for what? I was the primary instructor one week for 80 hours of day and night live fire training. We had very few NCOs so we had to pick the best and promote the best from our trainees. Then we received an assignment to U.S. Army, Pacific. Destination unknown. Let's see, it's the mid-60s, Pacific Command, destination unknown. Gee, I wonder where we're going."

Bob Babcock in Vietnam

July 21, 1966: "We were all home on leave when President Johnson announced that due to a nationwide airline strike, soldiers, airmen, and Marines who couldn't make their port call had nothing to worry about. He said for us to get there the best way we could. Out of 180 soldiers, only two didn't make it back on time. So, we go aboard the USNS Nelson M. Walker and reach Qui Nhon 16 days later, board a C-130 then fly to Pleiku in the central highlands. Engineer dump trucks were waiting for us. Now remember, we're a bunch of raw rookies, we had a couple Korean War veterans with us but none had experience in 'Nam. I mean we were green as grass. So we did like we were trained to do, got back to back in the dump trucks, then locked and loaded as we drove through Pleiku. Then we start looking around. There are G.I.s walking around, no weapons, no helmets, no web gear, chasing women and drinking beer. And we're thinking, 'this is a war-torn country?' Really weird."

"We were taken to what became the major 4th Infantry Division base camp, but at that time there were only a few Army tents. Then they told us, 'Okay, your perimeter is right over there, dig in, it's getting dark, and a patrol from the 1st Cav had spotted some NVA on a nearby hill.' Oh, hell, so we dig in and we're all scared, nobody knew what we were doing, but we were ready to do whatever we had to do. Then the monsoon hit, rain and rain and more rain, another company fired into the dark, flares were going off all over the place. We were advised that an attack was coming. Well, nothing happened. Come morning, we were a ragged bunch but were happy we'd made it through the night. For the next two weeks, we continued working on the base camp.

"I was assigned the first patrol of our division. I was given a little bubble chopper to reconnoiter with and once again I'm thinking, 'Oh, hell, we are going out there.' Two photographers went on patrol with us, one still photographer and one movie type guy. So, we're going along and these two guys go outside our patrol perimeter taking pictures, and I'm thinking, 'This is surreal.' I mean, we're in a combat zone, what is going on? I never saw the photos or film. I guess the two photo guys were disappointed because we didn't see any action."

Patrols, patrols, patrols. Aug. 27, 1966 through mid-October, 1966: Babcock and Bravo Company hook up with the 101st Airborne to begin patrolling along Highway 1 south of Tuy Hua. He said, "We patrolled an area called Mosquito Valley every day, all the time. We encountered a few VC and snipers. Then September 3, Charlie Company lost our first man KIA, PFC Albert Collins. Back at the 4th Infantry Division base camp, still under construction, the instructions were to name the base camp after the first trooper killed in action. Our Division commander, Major General Arthur Collins said no because the men will think we named the base after him. They decided to name the base camp after the first officer killed in combat. Well, we lost our first officer on November 5, a graduate of West Point class of '65, Lt. Richard Collins. General Collins then decided to name the base camp after the first posthumous recipient of the Silver Star. On December 2, 1966, Lt. Mark Enari was killed while fighting North Vietnamese regulars in the Central Highlands and posthumously awarded the Silver Star. Early in 1967, the 4th infantry Division's base camp at the foot of Dragon Mountain in the Central Highlands was named Camp Enari in honor of Lt. Mark Enari and kept that name as long as American forces were in Vietnam.

"We eventually ended up on the Cambodia border in Operation Paul Revere IV. From November 3, 1966 through December 22, we were in triple canopy jungle, I mean real jungle, making and hacking our way one to three clicks a day. We were looking for the bad guys. Well, we found them. Alpha Company found a lot of them, and got hurt bad. Charlie Company found a lot of them, and got hurt bad. Our men of Bravo Company did not get hit. A few small skirmishes, that's it, and I'll go to my grave wondering why we didn't get hit hard like Alpha and Charlie."

Bob Babcock with a captured AK-47

Merry Christmas: "Then in all their wisdom, the higher-ups call off the war for Christmas. We flew by Huey helicopters back to our base camp and I became the executive officer, the only officer who stayed in the same company the whole year. A new company commander arrived and on January 3, 1967, Bravo Company goes back to the same area and damn near got kicked out of there. During the two-week Christmas stand down the NVA, of course, came back in from Cambodia. We cleaned them out once, left, and they come back in...shows you how stupidly the war was handled."

On courage: "One of my squad leaders, Doug Muller from New York, absolutely hated the Army, but he was outstanding. When he got wounded, a medic named Julian White crawled out to help Doug, then he got hit. I was back at the fire base but met the chopper when they flew in. White was still working on Doug, it was touch and go. They also brought in the sniper that had shot them, but he was also in bad shape. We noticed White limping around. He said, "Well, I'm shot," but he kept working on Doug. Doug made it. The sniper died on the operating table."

Nowhere was safe: "It didn't matter where you went or where you bunked, you got hit by mortars; the enemy always knew where we were. One night our bunker took three direct hits...not a fun deal. But I came through it and came home in June, 1967, and decided to get out of the army. I got back to Oklahoma in early July. There weren't any protesters, but we were ignored. People would see me and ask, "Bob, where have you been?" When I replied, 'Vietnam', they would just say, 'Oh,' and change the subject. No thank you, kiss my foot, nothing."

Babcock joined the IBM Corporation and had a successful 34-year career in sales, marketing, and management. He joined the 4th Infantry Division Association in the early 1990's and during the last 20 years served as President for eight of those years. His accomplishments, his love of country, receiving a public service award by General Ray Odierno in front of 80,000 people, the author of five books, and C.E.O. of Deeds Publishing for the past 12 years would be a story worthy of its own book. And the motto of his publishing company: Deeds not Words.

His closing comments: "I was in firefights and skirmishes, but never a major battle. God also gave me the leadership ability and a loud booming voice that almost deaf veterans can hear, plus the desire to bring people together. Vets can heal each other. God had a mission for me after the war, bringing people together, so I'm doing more out of uniform than when I was in uniform. My wife, Jan, and I were in DC and I saw a man with a 4th Infantry Division patch rubbing a big section of the black granite on The Wall. I walked up to him and said, 'Tell me about it.' He did, the only member of his platoon to survive, and he did so by playing dead. That day my wife learned all about Vietnam."

Interviewing Bob Babcock is a history lesson best heard using a pair of earplugs. His thunderous voice commands attention, as does his grin, an obvious love for life, and a sense of humor sorely missing in today's society. His humility, however, concerning his service in Vietnam has the legitimacy of all combat veterans, of knowing he did his duties to the best of his abilities and relinquishes the glory to the names on a long black granite wall in Washington, DC.

His book "*What now, Lieutenant,*" reflects that humility, amid battle, heartbreak, death, and survival. But he shared this story. We all need to.

Hanging in a place of prominence in his home, visitors will discover his 4th Infantry Division patch, Combat Infantryman's Badge, a pair of Lieutenant Bars, Airborne Wings, and the Crossed

Rifles of an Infantryman, along with a plaque given to him by his children:

> For you Daddy,
> Because We think God each
> Day for the Freedom You Fought
> to give to us. We Love You, and
> are proud that you're our Father.
> *Krissie, Rob, and Mark*

A PROGENY OF AVIATORS

Arranging an interview quickly turns into something special when the veteran suggests, "Let's fire up the Stearman and fly down to Peach State Aerodrome for lunch at Barnstormer's Grill. After that we'll fly back and do the interview inside my hangar by the bar." No need to twist my arm.

The grass airfield in Williamson, GA is also home to the Candler Field Museum and Civil Air Patrol. Pilots who frequent Barnstormer's Grill are seasoned aviators with thousands of hours behind the controls of modern-day aircraft, but these guys would feel right at home inside the cockpit of a WWI era Sopwith Camel or Fokker Dreidecker (tri-plane) of Red Barron renown. Loiter around the grass field long enough and you may spot a floppy-eared Beagle land his dog house.

John Laughter can recite Barnstormer's menu by heart from

one too many hamburgers and tasty desserts shared with other daring young men and their flying machines. Thing is, many of the aviators, male and female, are no longer young and their flying machines are older than the pilots. Yet in aviation lingo 'no longer young' can usually be interpreted as an enormous amount of hard-earned flying experience by trial and error, sometimes even the death of a wingman. As a Navy fighter pilot, Laughter flew into the most heavily defended country of his generation, a place called North Vietnam. And this is his story.

Laughter: "I was born into a family of aviators in Horse Shoe, NC. My dad served with the 101st Airborne in WWII and was severely wounded on the second day of the Normandy invasion, which ended his military career. But the injuries didn't stop dad from becoming a private pilot, even with a steel pin in his leg. Dad was a great pilot; he soloed after only four hours of instruction. My mom conquered her fear of flying to become the second licensed female pilot in western North Carolina, and my uncle trained young cadets during WWII using the Stearman biplane. He became a commercial airline pilot and took me for my first flight in a Globe/Temco Swift. I fell in love with aviation and decided right then and there I wanted to be a fighter pilot, then later on fly commercial aircraft. I still remember going to the airfield with my family when I was 10 years old. I would sit forever in the cockpit of a Stearman duster. I bought that same Stearman in 1985."

Laughter's third year at East Carolina gave him the opportunity to fulfill his aviation dreams. He said, "Navy recruiters visited the campus and offered me a ride in a T-34 (Beechcraft Mentor).

Of course, I took them up on the invitation and after the ride I said to myself, 'Yeah, buddy, this is for me.' I signed up for cadet training and arrived in Pensacola in December of 1964. I figured the Navy had a plane waiting for me with my name on it but quickly found out things didn't work that way."

Laughter recalled the rebellious nature of his youth. "You know, after joining the Navy I finally realized everything my dad had taught me made sense. Dad was a disciplined man and he taught that discipline to me. It helped me become the aviator I am today."

Pre-flight School at Pensacola paved the way for primary flight school at Saufley Field. Laughter said, "I soloed in a T-34 and learned basic acrobatics, but I kept wondering if jets or props would be in my future. Luckily, I got jets. They sent me to Meridian, MS to master the T-2 (North American Buckeye)."

Once proficient behind the controls of a Buckeye, Laughter returned to Pensacola for gunnery training and carrier qual (qualification). He acquired his carrier landing and takeoff experience on the legendary USS *Lexington,* stationed in the Gulf of Mexico. Laughter's views on his first carrier landing: "I didn't think that much of it because of all the training the Navy had given me. Everything went as I had been directed. My first shot (catapult) off the carrier was perfect, a bit breathtaking, but perfect."

Laughter's next assignment: Chase Field in Beeville, TX to train on the F-9 Cougar and F-11 Grumman Tiger. One F-11 training mission tested his survival skills. "I was rolling in on target during gunnery practice when suddenly the 'low oil pressure' warning light came on, which meant 'get the heck home.' I called in my emergency and headed for Corpus Christi. So, I'm on a precautionary approach and let the gear down, then the darn plane falls out of the sky. Normal thrust failed, the gear is sucked back up, engine power was abnormal, and I'm just hoping for the best. I made it to the short runway but was forced to shut down the engine on final approach, then dead stick the aircraft back down to earth. I made it okay."

All the experience, all the training, all he ever learned behind the controls of an aircraft now came into play. Assigned to VF-124 squadron at Miramar near San Diego, Laughter mastered the F-8 Vought Crusader supersonic Navy fighter. "The F-8 was a great airplane, plus I married my forever sweetheart, the gorgeous Carole Bowden," he said with a smile." Newlywed Navy aviator John Laughter reported to the Sundowners of VF-111 and in April of 1967 set sail aboard the USS *Oriskany* for Westpac. Destination: Yankee Station in the Gulf of Tonkin off the coast of North Vietnam.

On Yankee Station: "We had seasoned pilots on the *Oriskany* who briefed us on their experiences the year before, but their information didn't help very much because of the new targets; the capitol city of Hanoi and the major port of Haiphong. I flew 'Iron Hand' combat escort missions for A-4 Skyhawks, several flak suppression missions using Zuni missiles, TarCap, Barcap, Migcap, and photo recon escort. During the first week, we lost one plane per day, then the losses skipped a day then we lost two the next day. My F-8 was the C model without hard points on the wing. We could carry four Zuni rockets or four Sidewinders. The E model Crusaders carried bombs because they had hard points on their wings."

On tactics: "A Skyhawk flying an Iron Hand mission would launch a Shrike missile at a SAM (surface to air missile) location. The Shrike would fly right down the SAM's radar beam and destroy the site, unless the operators saw it coming and quickly shut down their system. Our Crusaders would protect the A-4s plus escort strike forces into their target. We drew the SAMs toward us and away from the strike force. One tactic was to let loose a Zuni rocket toward Hanoi, which forced the SAM operators to shut down their radar because the Zuni looked like a Shrike coming at them. The North Vietnamese were armed with 37mm, 57mm, 85mm, and I think some 100mm guns plus SAMs. One of their tactics was to form a 'cone' of anti-aircraft fire. By doing that, the gunners forced our planes to maneuver through a barrage of lead. If we missed the target we'd have to go back another day and fly through the same 'cone' again."

On losing a pilot: "That was tough. You're sitting in a briefing room with 30 guys and you know at least one of us most likely won't come back. You just hope it's not your day to be that pilot. On our first cruise, the A-4 Skyhawks got a real pounding, we had 28 planes lost or damaged. We utilized a tactic called 'wet-winging' to help improve our survival rate. If you're shot up and leaking fuel and lucky enough to reach the sea, then an aerial fuel tanker would hook up to you and keep pumping fuel into your plane to replace the fuel being leaked. Some holes in our planes were large enough to stick your head through."

John flying inverted - fuel leaking from rear is 'venting' when inverted

The cameraman: "One of our pilots, a character named Stattin, used his own 8mm camera to film missions. We nicknamed the guy Cecil B. DeStattin. On one mission, Stattin was filming SAMs heading our way when he realized a SAM was heading right for him. He broke hard to avoid the missile but the 400lb warhead exploded and peppered his aircraft with shrapnel. He made it back with over 100 holes in his plane.

"We continued to hit the enemy hard and did not back down because of losses. We did our best under the circumstances and we did it for our wingman and our squadron. The anti-war protesters back in the States didn't bother us at all." Their tour completed, the *Oriskany* headed for Hong Kong for a little holiday cheer during

Christmas. "Well, that didn't last long," Laughter recalled. "The replacement carrier was delayed so we had to go back on station for 10 days in January of '68. Those 10 days cost us two more planes and two more pilots. Thus ended the bloodiest combat cruise of the Vietnam War."

Note: Up against enemy Mig fighters, the Crusaders earned the nicknames "Mig Master" and "Last of the Gunfighters." Migs did not often tangle with the Crusaders since the enemy pilots knew they would probably lose the engagement. The final war tally for the F-8s: 19-3 in favor of the Crusaders, the best of the Vietnam War. On Laughter's first cruise, the F-8s only had one Mig engagement: six Migs against one Crusader, ending with one Mig shot down.

Returning home to San Diego, a reserve F-8 squadron called up by President Johnson took over their fighter planes. "Well, that didn't last too long either," Laughter stated. "Most, if not all, of the reserve pilots were commercial airline pilots. They took over the squadron's airplanes but bellyached about everything, including their loss of pay. Then it quickly became obvious these guys couldn't land on carriers at night. The Navy sent them home then gave us back our airplanes."

Laughter recalled an important incident during his first tour. "I forgot to tell you about the USS *Forrestal* disaster. Our sister carrier was launching planes from the VF-11 Squadron on July 29, 1967 when a Zuni missile misfired on the flight deck. The wayward Zuni struck a fully fueled and armed A-4 piloted by the future Senator and Presidential candidate John McCain. It was an unbelievable disaster. The resulting explosions and fire killed 134 men, injured another 161, and destroyed 21 aircraft. The pilots and aircraft that survived the conflagration were sent over to our carrier, the *Oriskany*, so the support missions could continue."

Laughter recalled John McCain's last mission: "In late October, we hit the thermal power plant in Hanoi. It was on that mission that McCain went down and became a POW. One thing that's interesting yet depressing; our newest weapon at that time was the Walleye (first in a series of 'smart' bombs) that our pilots knew very little about. But the North Vietnamese had all the information they needed on the Walleye and knew how to jam or evade the weapon. That meant we had a traitor back in the States. And let me mention Jane Fonda. That picture of her sitting in the seat of a North Vietnamese anti-aircraft gun irked all of us; it was hard to swallow. We were on the receiving end of that gun and Fonda's publicity stunts only encouraged our enemies. Most of us still hold a grudge about that incident but we don't dwell on it."

Laughter's second tour on Yankee Station wasn't a cake walk, but at least offered less of a risk. He said, "President Johnson had initiated another bombing halt to entice North Vietnam to accept a solution to the war. The bombing halt took Hanoi and Haiphong plus other important targets off the bombing list. We still had losses, but not quite as severe as the first tour of duty. We did our job; we served our country, and we were proud of our service."

John Laughter flew 155 combat missions into North Vietnam. His Crusader was never hit, not one single scratch. "Yep, I was lucky, and I know that," Laughter said, grinning. "My Guardian Angel earned a lot of overtime pay."

John next to his Crusader

How good is a combat experienced Navy pilot? Laughter received high marks at the first Top Gun school, helped develop new tactics to fight enemy Mig fighters, transferred to the East Coast and flew F-4 Phantom Jets off the restored USS *Forrestal* as a member of the renowned VF-11 'Red Ripper' Squadron, and served as a weapons training officer.

On hitting the silk: "I was in test squadron VX-4 at Point Magu, CA. We normally used the Pacific Missile Range as our primary operating area. One day while I was maneuvering behind an F-4 Phantom in a simulated dogfight, my F-8 went out of control and rolled to the right. One of the connecting rods had broken. I regained some control but could only perform right turns. I struggled with the aircraft for about 15 minutes, then I knew it was time to eject. We were over a sparsely populated desert so I maneuvered toward some barren mountains. I pulled the handle, then the rocket seat ejected me out of the aircraft. It would take me a lifetime to explain my thoughts during the few seconds before the chute opened, but I can tell you one thing, when that orange and white parachute deployed it was without a doubt the most beautiful thing I'd ever seen!" After all the careful planning and piloting, the F-8 still landed in a civilian's backyard. Laughter recalled, "The jet just spun in. No injuries occurred on the ground but there was not one piece of the jet you couldn't pick up in your hand." The Crusader's engine was found buried 28 feet into the ground.

After almost nine years piloting Navy jets, Laughter decided on a career in commercial aviation. He retired in 2003. Asked how many different aircraft he'd flown, Laughter replied, "It would be easier for me to name the planes I haven't flown."

His final thoughts on the Vietnam War: "Vietnam was a terrible war that was terribly managed. We lost a lot of good people, then in the end we gave it all away. For me, it was a good time in my life, learning who I am, what I am, and operating under pressure. But you know, we have heard 'no more Vietnams, never again,' yet we see them doing the same thing over and over and losing everything we've gained. I just don't get it."

"The duty of the fighter pilot is to patrol his area of the sky, and to shoot down any enemy flying in that area. Anything else is rubbish."

—Baron Manfred von Richthofen, the 'Red Baron'

BASEBALL, BRAVERY, BLEAKNESS, & BORN-AGAIN

All of us have dreams, especially youthful dreams. In high school, we can't wait to get out; then after joining the Rat Race, we wish we could get back in. In the real world, dreams are shattered by pragmatism and survival. Few ever make rock star status or sashay down the runway as a super model, even less make an honest politician or a Mother Teresa.

All of us have dreams. Joe May dreamed of playing professional baseball. Joe ate and slept the game. Baseball was his life. He liked the grass stains on his uniform, the band of sweat circling his cap, the feel of his Louisville Slugger, the dust of the infield, the crack of the bat from a solid hit. After college, the Pittsburgh Pirates gave Joe the opportunity to achieve his dream.

There is no doubt a certain amount of courage is required to face a 90mph fastball. At that speed, in a contest between face bones and a baseball, face bones lose. Nevertheless, Joe possessed the professional bravado to gleefully step up to the plate. He would need that strength of will: Within six months, he'd be drafted into the United States Army. They, too, needed courageous men, in Vietnam.

Joe May in Vietnam

Joe May answered his call to duty. He served his country. He fought in a war. Joe witnessed the dreadfulness of war, tasted war, smelled war, and learned to hate war. His Field of Dreams became lifelong nightmares. He would never again be the same. Yes, Joe fought the good fight and obeyed confused if not questionable orders. He burned villages; he lost friends; and for many years to come, Joe lost himself.

Joe May inside crater caused by VC 122mm rocket

Hollywood has it all wrong. There is no glory in war. But there is suffering, persistence, and getting the job done; trusting your buddy, saving your buddy, your buddy saving you. It requires teamwork, just like the game of baseball. True, glory exists in baseball, but rarely does a player achieve glory without his teammates. In times of war, 'glory' is often documented then recognized by upper echelons, but not by the lowly grunts. Their Medal of Honor or Silver Star or Purple Heart originates from self-sacrificing conduct on a battlefield; no supermen, no glory-hounds, just soldiers protecting or supporting or saving other soldiers. Then the soldiers come home.

Mothers and fathers, wives and sweethearts, brother and sisters, an aunt or an uncle, offer kisses, hugs, a handshake, or perhaps a kindly 'welcome home' to the returning veteran. Any welcome is welcomed. But the soldier who left is not the soldier who returned. The change is apparent. A relative may attempt to recognize the trauma by quietly stating to another relative, "He's not the same," then both relatives move on with their lives.

Joe had changed, exceedingly so. Nightmares became the norm. Coping as a citizen became a daily struggle as if he had another battle to fight, another long year to survive. As months became years, the adjustment became an excruciating struggle, yet Joe was helping others instead of getting help for himself. He was lost; he was hurting, and hurting the people who loved him.

Joe finally received the help he needed. He got his life back on course and realized his torment was not a lonely journey; there were others, countless others, suffering the same pain, the same agony, having the same nightmares, having trouble coping.

Born-again doesn't always mean a 'hallelujah-I'm-saved-struck-by-lightning' conversion with God. In Joe's case, being born-again means into a productive life, getting back on course, and helping others get help. Joe is back with us.

Welcome back, Joe, and welcome home.

* * * * *

Professional Gospel singers Joe and Viola May welcomed their son into this world in 1945. Natives of St. Louis, Missouri, the Mays and their seven offspring eventually moved across the muddy Mississippi to East St. Louis, Illinois where young Joe completed his schooling. Joe's baseball talent at Lincoln High School soon caught the attention of collegiate baseball scouts. In 1962, Joe received a baseball scholarship to Tennessee State University.

At Tennessee State, Joe played baseball under the vigilant eyes of professional baseball scouts. After graduating in 1966 with a B.S. Degree in Sociology, Joe happily signed a contract with the Pittsburgh Pirates. He stated, "My signing bonus was $5,000.00. Shoot, that was a lot of money in 1966."

Six months into his long-dreamed of baseball career with the Pittsburgh Pirates' Class A minor league team in Scranton, PA, Uncle Sam came calling. Joe was drafted into the United States Army. He said, "The war in Vietnam was escalating and as a single man without dependents I knew my goose was pretty well cooked."

Basic training took place at Ft. Leonard Wood, MO. There, the Army conned Joe into signing up for three years instead of the normal two-year enlistment. He recalled, "They said if I signed up for a three-year enlistment then I could sidestep Vietnam, so I signed on the dotted line."

His next assignment was Fort Gordon, GA for advanced infantry training as a communications specialist. Completing the course, Joe and his fellow infantrymen were roused from their bunks for an 0600 formation. Joe said, "They called out the names of guys going to Korea or Germany, then said, 'the rest of you men will be going to Vietnam.' My name wasn't among the ones going to Fort Huachuca or Korea or Germany." Apparently, the Army had pulled a con job; Joe May was going to war.

Joe stated, "I cried, I pitched a fit, I tried everything. I went to church, got saved, nothing helped. I was flown straight to Ton Son Nhut Air Force Base in Saigon, Vietnam, then flew on to Pleiku in the Central Highlands."

He hooked up with the 4th Infantry Division and was assigned to the 245th Psy Ops Unit (Psychological Operations). UH1-D Huey choppers became his airborne home. He said, "I'd fly out to different villages with three or four other guys and an interpreter. We'd set up our speakers about a hundred yards from a village then try to call the villagers out. Shoot, nobody came out except children and old women."

Joe quickly learned the tricks of the trade, like the Viet Cong soldiers taking refuge in underground caves and tunnels. "Shoot, they would hide anywhere," he said. "After we cleared a village of inhabitants, the mission turned into a Search and Destroy campaign. We received orders to level the villages. I remember on this one mission we were torching the huts and I tossed a hand grenade into a flimsy old hootch. I'll regret that forever. A woman and kid came out screaming with horrible, disfiguring wounds."

The incident traumatized Joe. He said softly, "I could not forget that sight." The platoon leader told Joe he had two choices: toss in grenades or stick his head inside the primitive huts, in which case the VC would probably blow his head off. Joe spent about five months with the Psy Ops Unit, usually dropping propaganda leaflets from choppers into VC controlled territory. After Psy Ops, Joe transferred to the 3rd Brigade, 25th Infantry Division.

Joe recalled, "Our LZs (landing zones) were hacked out of the jungle. I was just regular infantry by then, armed with an M-16, seven or eight grenades, and machine gun belts wrapped around my torso. I looked like Pancho Villa." The stagnant jungle heat took a heavy toll on the physical element, but war can also take a heavy toll on the mental component. Joe's platoon leader was decapitated from a direct hit. "I was standing right next to him when he got hit," Joe said, biting his lower lip. "I lost another friend. Folks just don't understand what war is."

Recalling the B-52 strikes, Joe said, "Oh, man, the earth would shake like Jell-O. Mountains surrounding the base would erupt in smoke and fire and leave nothing but bomb craters. Watching a B-52 strike makes you glad you're on the giving end, not the receiving."

During the countrywide Communist Tet Offensive of'68, Joe's base camp was pounded mercilessly by rockets and mortars. "The rockets and mortars usually preceded an enemy ground assault," he recalled. "The Russian-made 122mm rockets made craters the size of a school bus. After the barrage is over, here come the communist troops charging over the barbed wire. We'd mow them down like wheat but it didn't matter, we still lost too many of our own people."

Joe participated in an U.S. counter-strike in the A Shau Valley. "As soon as the choppers touched down, we were ambushed by North Vietnamese troops concealed in the jungle. They cut us to pieces. Within two days, over two hundred guys were killed and hundreds more were wounded. I can still hear those boys calling for their mommas."

Joe made it home. "I remember some hippie-like people calling me a baby-killer. That hurt. We never intentionally killed kids, but the kids sure as heck killed us." Joe mentioned how the Viet Cong would wire small children with explosives, and Coca-Cola bombs. "You'd see a small child in a village, maybe five or six years old, holding a Coke can approaching a G.I., then all of a sudden ... boom! And that was that."

Returning to Tennessee State, civilian Joe May took advantage of the G.I. bill and earned a Master's Degree in Counseling and Psychology. He also met his wife, Pat, a champion track star at Tennessee State. Joe said, "We were married for 13 years. I was making a living at the VA as one of their Mental Health and Readjustment Counselors....can you imagine that? I was the one having the nightmares and cold sweats, fighting in my sleep, horrible flashbacks, a fear of big crowds, anger over little things; yeah, it took a toll on our marriage. I was the soldier in need of counseling; I was a time bomb."

In 1995, after years of secret torment with too many memories and not enough rest, the VA diagnosed Joe 100% disabled from PTSD. "I eventually came to Atlanta," he said. "I found work at Families First and continued my meetings with PTSD group sessions. The guys are like me, we relate, we save each other." Now a lifetime member and spokesman for Disabled American Veterans, Joe helps his fellow soldiers receive the assistance they require, indeed, the assistance they have earned. Joe said, "The walking wounded are still with us, homeless, wandering the streets and needing help. I'm dedicated to helping them; they are my lost Band of Brothers."

Reunited with his wife Patricia in 2009, Joe was diagnosed with ischemic heart disease due to exposure from the herbicide Agent Orange while serving in Nam. Fully recovered from open heart surgery, he stated in his closing comments, "You know, when I finally leave this world I hope people remember me as a person that loved everybody. I learned in war that we are all humans. There has got to be a better way to resolve differences other than war."

ONE SOLDIER'S MEDAL OF HONOR

Frequently referred to as the Congressional Medal of Honor, our nation's highest award for gallantry above and beyond the call of duty is officially called the Medal of Honor. Decorations for bravery originated in the American Revolutionary War after three militiamen from New York State detained a British spy, thus saving West Point from capture by the British. All three received a decoration called The Fidelity Medallion.

In 1782, George Washington established the earliest formal system for awarding acts of uncommon valor: The Badge of Military Merit. During the two-year Mexican-American War, the bravest of the brave were awarded a Certificate of Merit. The award, however, was discontinued after the conflict, only to be re-introduced in 1876.

Astonishingly, there were no standard awards or medals available at the start of the American Civil War. Finally, on December 21, 1861, President Abe Lincoln signed into law the Navy Medal of Honor. An act of Congress in 1863 made the Medal of Honor a permanent award, but for enlisted men only. Army officers became eligible in March of 1863.

On March 3, 1915, the Medal of Honor finally became available to officers in the Navy, Marines, and Coast Guard. In 1963, a separate Coast Guard medal was authorized, followed by an Air Force Medal of Honor officially adopted in 1965. As of this writing, the Coast Guard Medal of Honor has yet to be designed.

Nonetheless, in the beginning, sloppy requirements warranted the Medal of Honor being awarded for 'courageous' acts of bravery not related to combat. All that changed in 1918 with an enhancement of the requirements which stated the Medal of Honor could only be awarded for bravery against a recognized enemy. The new guidelines struck 911 names from the Medal of Honor Roll, including soldiers of the 27th Maine who received the Medal in the Civil War simply because they reenlisted. Another 29 soldiers whose only accomplishment was an assignment to Lincoln's funeral detail, also had their Medals rescinded. Buffalo Bill Cody was among the six civilians struck from the Roll.

Tougher present day requirements now include *"conspicuous gallantry and intrepidity at the risk of life above and beyond the call of duty."* Members of the military are urged to salute Medal of

Honor recipients as an act of courtesy and respect, no matter what rank the recipients may hold. Even so, there are no edicts or regulations requiring the courtesy salute.

Roy Benavidez, MOH recipient

Less than 3,500 military personnel have received the Medal of Honor; only 19 have been awarded two. Notwithstanding, the requirement "*actions against a known enemy*" was ignored for Charles

Lindbergh and Major General Adolphus Greely. The Medal has been awarded to the 'Unknown Warriors' of foreign countries, including England, France, Romania, Belgium, and Italy. In the United States, Unknowns received the Medal of Honor in WWI, WWII, Korea, and Vietnam. The Unknown from Vietnam, 1st Lt. Michael J. Blassis, USAF, was finally identified by DNA procedures. DNA identifying techniques have made future Unknowns highly unlikely.

The most Medals of Honor were awarded during the Civil War: 1,522. The least number awarded for a conflict was one, given to Capt. William McGonagle for his gallantry during the mistaken identity event when Israeli planes and torpedo boats attacked the USS Liberty in the Six Day War in 1967. The attack killed 34 Americans and wounded another 171.

In contemporary America, the Medal of Honor represents the courage and selfless behavior under extreme circumstances by the men and women of our armed forces. And to be candid yet compassionate, the Medal of Honor usually means one human life sacrificed to save the lives of many others.

Army Sgt. Raul (Roy) Perez Benavidez served his country in Vietnam and saved 'the lives of many others' with no regard for his own life. The fact that Sgt. Benavidez lived to personally accept the award from President Ronald Reagan is a miracle in itself.

RAUL (ROY) PEREZ BENAVIDEZ
AN AMERICAN HERO

Shortly before President Reagan awarded the Medal of Honor to Sgt. Raul "Roy" Perez Benavidez, the Commander-in-Chief turned around and stated to the press, *"If the story of his heroism was a movie script, you would not believe it."*

Roy Benavidez was born in 1935 near Cuero, TX to poverty-stricken sharecroppers of Mexican and Yaqui Indian ancestry. Both parents died of tuberculosis before his eighth birthday. He and his younger brother, Roger, along with eight cousins, were raised by a grandfather and an aunt and uncle in El Campo, TX.

As a young man growing up in humble surroundings, Benavidez shined shoes at the local bus station, labored on farms in Texas and Colorado, and worked in a tire shop. Racial slurs like, 'the dumb Mexican' were common. Seeking a better life and better opportunity, Benavidez dropped out of school in the 7th grade.

In 1952, he enlisted in the Texas Army National Guard and in 1955 entered the regular Army. Benavidez won the hand of Hilaria "Lala" Coy and the couple wed in 1959, the same year he finished airborne training and joined the 82nd Airborne Division.

Deployed to Vietnam as an advisor in 1965, Benavidez stepped on a land mine while on patrol and received a host of injuries, specifically back injuries. He was evacuated to the United States for extended treatment at Brooke Army Medical Center where the physicians claimed he would never walk again. Benavidez didn't listen. Despite his spinal injuries, Benavidez walked out of Brooke

Army Medical Center in July of 1966. His courage and faith had served him well. It would not be the last time.

Roy Benavidez at 15-years-old

He returned to active service and trained at Fort Bragg, NC for the elite Studies and Observations Group. Still suffering from unrelenting back pain, Benavidez returned for a second tour of

Vietnam in January of 1968. He was assigned to Detachment B56, 5th Special Forces Group Airborne; the 1st Special Forces at Loc Ninh.

A devout Catholic, Benavidez was attending prayer services on May 22, 1968 when he heard a desperate plea on the radio, *"Get us out of here! For God's sake, get us out!"* The frantic call for help came from a twelve-man Special Forces Recon Team pinned down in thick jungle and surrounded by an NVA (North Vietnamese Army) regiment west of Loc Ninh. Three choppers had already attempted a rescue but were driven back by small arms and anti-aircraft fire. Benavidez didn't wait for orders or permission. With a medical bag in one hand and only a knife for protection, he jumped into the bay of a Huey revving up for another rescue attempt.

THE MAKING OF A LEGEND

Intense enemy fire kept the chopper from landing. As the Huey hovered 10 feet off the ground, Benavidez made the sign of the cross across his chest, then leaped out of the chopper. The Injured recon soldiers lay 75 yards away. Benavidez started a deadly gauntlet, fell when an AK-47 round pierced his right leg, jumped back up only to be knocked back down by a grenade explosion that ripped flesh from his back and neck. Those initial injuries were the first of 37 separate bayonet, shrapnel, and bullet wounds he would receive during the next six hours of fierce combat.

Praying aloud, Benavidez mustered the courage to rise again

and ran under fire to the crippled 12-man squad. He found four soldiers dead and the other eight badly wounded. After distributing ammunition to the soldiers still able to fight, Benavidez injected morphine into the wounded while calling in air strikes. He was suddenly hit again. Bleeding and in pain, he dragged the dead and injured men to the hovering Huey while providing cover fire with an assault rifle he found on the ground. The Huey moved to recover even more bodies; enemy fire increased.

Benavidez made another run to retrieve classified documents of radio codes and call signals tucked away on the body of the dead team leader. Seizing the documents, another round pierced his stomach and more shrapnel peppered his back. Coughing up blood and ignoring horrific pain, he tried to make it back to the chopper only to witness the pilot receive a mortal wound. The Huey spun, then crashed to the ground.

Seemingly unstoppable, Benavidez began pulling the wounded from the overturned chopper as he called in more airstrikes and directed helicopter gunship fire. Blood trickled into his eyes, he lost vision, waited a few seconds for his sight to clear, then set up a defensive perimeter and rallied the injured soldiers to fight on. He later recalled, "I made the sign of the cross across my chest so often my arms looked like an airplane propeller."

Before a second Huey arrived, Benavidez was wounded several more times, yet continued to urge the injured to fight on, and to keep praying. As the chopper approached the area, Benavidez slung a wounded soldier over his shoulder and staggered toward the LZ. An enemy soldier suddenly jumped up and clubbed Benavidez from behind with the butt of a rifle. Still clutching the injured soldier on his shoulder, Benavidez fell to the ground. Bayoneted in both arms, he grasped the bayonet, a tactic that saved his life but gouged his hand. Grasping the bayonet allowed enough time for him to retrieve his knife and kill the enemy soldier. His jaw shattered, his arms lacerated, his hand deeply slashed, bullets and shrapnel had pierced his body. But Benavidez still retained enough strength to rescue one more injured soldier, their Vietnamese interpreter.

The wounded soldiers in the chopper who were physically able pulled Benavidez's battered body aboard the Huey. At first his blood pooled, then flowed out the open bay. Benavidez held his intestines in his hands during the 20-minute flight back to Loc Ninh. At the base, he was triaged then pronounced dead. As a surgeon attempted to zip up the body bag, the intrepid soldier marshalled enough strength to do one thing: he spit in the doctor's face.

The six-hour battlefield heroics of 32 year old Roy Benavidez was one of the most astonishing feats of the Vietnam War. He was hospitalized for over a year, yet refused to accept praise for his heroics, claiming, "No, that was my duty." His commander assumed Benavidez would not survive long enough to receive the Medal of Honor due to an extensive processing procedure for MOH consideration. As a result, Roy Benavidez was only recommended for the Distinguished Service Cross. Years would pass before the wrong could be corrected.

On February 24, 1981, President Ronald Reagan presented

Raul "Roy" Perez Benavidez the Medal of Honor. He accepted the nation's highest military award with two pieces of shrapnel still lodged in his heart and a punctured lung.

Benavidez receiving MOH from President Reagan

Benavidez's advice to young people: "An education is the key to success. Bad habits and bad company will ruin you."

His fatherly guidance to his son, Noel: "Never bring shame on our family name.

His three children are college graduates. In an unprecedented honor by the United States Navy, the Bob Hope-class roll on-roll off vehicle cargo ship, the USNS *Benavidez*, is named for an Army sergeant.

Scuttlebutt has it that when Special Forces troops are in a tough scrap and things are going badly, if additional courage needs to be summoned, they use the radio call sign *Tango Mike Mike*, Roy Benavidez's call sign.

Medal of Honor recipient Roy Benavidez died on November 29, 1998 from respiratory failure and complications of diabetes. He was 63 years of age. He was buried with full military honors at Fort Sam Houston National Cemetery; his Medal of Honor is on display at the Reagan Library in Simi Valley, CA.

YOUR LOVING SON

"Come on, you sons of bitches—do you want to live forever?"
—Gunny Sgt. Daniel Daly, USMC, WWI, June 1918, Belleau Wood, Two-time Medal of Honor recipient

Marines are trained to fight, to kill; to win. A Marine cook can slice a hunk of roast for your dinner plate or slice your throat, he's an expert at both. A Marine pilot can put a missile up your nose or your choice of orifices from inside the cockpit of his Harrier or blow your brains out with a handgun, he's an expert at both. A Marine lawyer can cite military law or lay down the law or if necessary exterminate an enemy combatant. He, too, can call upon deadly skills few people ever learn.

A Marine is a Marine, from the recruit to the Commandant; and once a Marine, always a Marine. These warriors are physically and mentally resilient with an aura of invincibility in combat yet can quickly convert into a messenger of kindness and respect for their fellow human beings. The key word is *Discipline.*

On March 20, 1779, Marine Captain William Jones advertised in the *Providence Gazette* for 'a few good men' to make a voyage aboard the Continental ship *Providence.* Ever since publishing that advertisement, the United State Marine Corps has continued to pick 'a few good men' to become the standard bearers for an elite group of fighters steeped in tradition, dignity, and honor. I have known and interviewed more than just 'a few' Marines, but they are all 'good men.'

THE PROPOSAL

I received an e-mail from my editor asking that I contact an organization called the Atlanta Vietnam Veterans Business Association, better known as the AVVBA. The group requested that a local columnist assist them in promoting a memorial event scheduled at the Walk of Heroes Veterans War Memorial in NE Rockdale County, GA. I jumped at the opportunity.

I e-mailed the AVVBA with my proposal. They liked it. Thus began a two month mission promoting a memorial service for a young Marine chosen to be honored that year by the AVVBA. I considered the assignment as an occasion to make new contacts and was elated when informed that the article would be featured on the front page of the East Metro section of the Sunday paper. Ideas flourished on the prospects of additional readership and media exposure.

However, it wasn't long before I understood my self-centered acumen was inappropriate and embarrassing. As I began my research and discovered new comradeship in the long list of interview candidates, I increasingly understood that as an Air Force flyboy I'd be interviewing Marines with hands-on combat experience, Marines with psychological and physical wounds, Marines devoted to the memory of fallen comrades, Marines that expected things to be done right.

Britt and buddies at Khe Sanh, Britt second from left

I started second-guessing my writing skills. The young Marine to be honored lost his life during the historic Battle of Khe Sanh. So how could I, a typically well-fed and clean shaven Air Force veteran of Nam, usually with access to a cold beer or a hot woman or both, write a narrative paying tribute to a US Marine who lived in the mud and dirt, ate cold food, patrolled some of the most dangerous real estate on earth in 1968, dodged rockets and mortars day and night, and participated in one of the few battles in Vietnam when Leathernecks were ordered to 'fix bayonets.'

To use the term 'honored to write the article' doesn't even give appropriate reverence for being offered the privilege to do so. Nevertheless, 'panicky' may be a better term for the emotions that fogged my creativity as the graphic specifics of the bloody mêlée that morning at Khe Sanh was articulated by the battle-hardened Marines who fought there. I had to face facts, that, as a flyboy, writing about Marines was an endeavor best reserved for a Marine historian. I decided to axe the opportunity until an easy-going member of the AVVBA, Lt. Col. Jon Bird, offered a little thoughtful advice on how to write the story. He suggested, "Think like a Marine."

So I tried; I gave it my best shot. But I'm not a Marine, and the Marines that put the event together knew I wasn't a Marine. I didn't fit in. I felt uncomfortable and lacking the needed skill by my own insecurities. But then again, all that didn't matter. The Marines chose me to do the job; the Marines placed their trust in me to do the suitable thing; the Marines asked me to honor a fallen Marine as he should be honored; the Marines invited me to the program rehearsal to acquire more appreciation of and about what it means to be a Marine. As I observed and learned from the Marines, I finally understood my necessity to muster the same resolve as they do in combat with the simple words, "Follow me." For two months, I almost felt like a Marine.

THE NEWSPAPER TRIBUTE

A Memorial Service will be held at the Walk of Heroes War Memorial on the grounds of Black Shoals Park in the northeast corner of Rockdale County at 11:00 am on Thursday, May 26. The Marine to be honored is Silver Star recipient PFC Ted Dennis Britt, killed in action on March 30, 1968 during the battle at Khe Sanh, Vietnam. A graduate of Southwest DeKalb High School, he was nineteen years old.

The ceremony, sponsored by the Atlanta Vietnam Veterans Business Association, will feature bag pipes and a Marine band. A USMC Color Guard will present the colors plus an UH-1 Huey helicopter and a C-130 Hercules transport will perform flyovers. The dedication address will be presented by Medal of Honor Recipient Col. Harvey "Barney" Barnum, USMC, retired. The Britt family and honored guests will witness the Folding of the Flag Ceremony after which the flag will be presented to PFC Britt's mother and his kid brother, Brigadier General Tim Britt, ret.

FIX BAYONETS

Pfc. Ted Britt arrived "in-country" on December 15, 1967. This young Marine knew, as did every Leatherneck at Khe Sanh, the President of the United States, even local tribal chieftains, that something big was in the making.

The timeworn French airfield on the Khe Sanh plateau in the northwest corner of South Vietnam was taken over from Army Special Forces by the United States Marines in 1966. The Leathernecks quickly lengthened the runway and strengthened their defensive positions, giving them the much-needed staging area and fortified strong point to conduct spoiling attacks against the North Vietnamese Army (NVA) infiltrating across the DMZ into South Vietnam. Itching for their own fight, an estimated 40,000 battle-hardened NVA soldiers laid siege to the 6,000 Leatherneck contingent on January 20, 1968. From that ill-fated day, the entire world realized the siege was on.

The NVA lobbed in as many as 1,300 mortar or artillery rounds on a single day at the Marines, who simply hunkered down and took the best the NVA could throw at them. Artillery duels became daily events. The enemy dug trenches within a hundred meters of the barbed wire and tested the defenses with ground probes. The Marines held their ground, waiting for opportunities, then would counter-attack in what became free-for-all engagements. Casualties mounted; reinforcements poured in, on both sides.

On February 23, Khe Sanh's ammo dump took a direct artillery hit. Over 1,500 American artillery rounds blew sky-high with the resulting detonations continually rocking the entire base. American air power pulverized the surrounding hills into a No Man's Land moonscape even as NVA anti-aircraft fire took an appalling toll on the supporting aircraft. Then on February 25, a Marine platoon conducting a perimeter sweep stumbled into an ambush and was annihilated, almost to the man. Unavoidably, Marine dead were left on the battlefield; Marines do not leave Marines on a battlefield; yet Marine dead had been deserted, and the Marines were determined to get their brothers back.

Pfc. Ted Britt and the Marines of B Company, 1st Battalion, 26th Marine Regiment, 3rd Marine Division, were roused from sleep around 3:00am on the smoggy morning of March 30, 1968. Moving carefully outside the base perimeter, the detachment paused until the supporting artillery barrage lifted. As it lifted, an unnerving graveyard silence encompassed the thick fog bank, no sound, no nothing. Tense moments passed. Then somebody heard the clicks of metal scraping and muffled voices speaking Vietnamese: there was no doubt weapons were being armed and the voices belonged to NVA soldiers. Shadowy movement was noticed within the fog bank, stealthy, eerie, but detectable. In whispers, the order filtered down from the Captain to the platoon leaders to squad leaders to fire support leaders: 'Fix Bayonets!'

As if caressed by the hand of God, the fog gently lifted. Into the jaws of lethal metal the Marines charged headlong into the lineal descendants of the North Vietnamese Army's 8th Battalion, 66th Regiment—the infamous 304th Iron Division, the same division that defeated the French at Dien Bien Phu in 1954. It has been said that payback is hell.

Marine Pfc. Ted Britt

For four hours, soldiers fought hand to hand, man against man, grenade against grenade; steel against steel, a knock-down drag-out barroom brawl for which the United States Marine Corps throughout their rich history are known to be most proficient. Many of the Marines engaged in the bloody battle were rookie replacements up against the best the NVA could field. But the NVA's best was simply not good enough.

Pfc Britt's squad was abruptly pinned down by machine gun fire and mortars. Aware that his squad faced imminent annihilation, Britt identified the main threat then charged an enemy automatic weapons emplacement. He singlehandedly destroyed the enemy emplacement, killing four NVA soldiers in the process. With the threat neutralized, his squad moved out. Within moments another enemy position opened up with lethal fire that held the Marines at bay. Courageously, Pfc. Britt launched another single-handed attack against entrenched NVA. While delivering accurate fire against an NVA fighting hole, Pfc Britt was fatally wounded.

The Leathernecks won their barroom brawl and decisively so. Casualties for the Marines numbered less than 40, yet over 150 NVA soldiers were killed, including their battalion commander and his entire staff.

VOICES

Peter Weiss, PFC Britt's platoon leader: "I didn't personally see Ted's fearless heroics but I heard about it from other platoon members. I was a 24-year-old lieutenant leading seventeen and eighteen year old Marines into combat for their first time. I was so proud of them. Those young Marines proved their courage in hand to hand combat. I think the order to 'Fix Bayonets' came from Captain Pipes, so we understood we would be mixing it up at close quarters with the North Vietnamese. We lost an entire platoon a month earlier, so the order to 'fix bayonets' provided us with a psychological 'edge' as we moved forward."

Mike McCauley, Ted's best friend at Khe Sanh: "I landed 'in-country' during November of 1967. Ted and I met on Hill 881 overlooking Khe Sanh. My Boston accent and Ted's southern drawl formed a quick friendship, me being the damn Yankee and Ted being the southern Rebel. I remember Ted always got fistfuls of letters from home when the choppers dropped the red mail bag. He would spend hours reading every word. We were jealous he got so much mail.

"During one shelling, our ammo dump took a direct hit. Live munitions were flying all over the place. A lot of our guys were injured when unexploded shells struck them. We would shell the NVA then they would shell us right back. We loved the Air Force. When the B-52s dropped their huge payloads, we would jump up and down cheering. It was awesome.

"In February, we lost about 25 Marines in an ambush. We were on the line and could overhear the radios crackling with guys screaming back and forth; we actually heard what was going on but couldn't do a damn thing about it. And we couldn't retrieve our dead...that was tough to take.

"What we experienced is like it happened yesterday. On March 30, we were nudged awake around 0300 then silently worked our way outside the base perimeter. There we paused and waited in dense fog for the artillery barrage to lift. I was squad leader. At daybreak, Lieutenant Weiss's voice suddenly cracked over the radio, 'Mac, tell your squad to fix bayonets.' I kept thinking, '*Oh, Jesus, this is scary,*' but I told my squad to 'fix bayonets'. There was a lot of nervous laughter, until the artillery stopped. Instead of the incoming we were surrounded by a dead calm, it was weird, like an omen. Then the fog lifted. Little by little, the dead calm became a noise like popcorn in a microwave. Pop. Pop...pop. Pop...pop... pop...pop, pop, pop, pop, that's how it got started, then all hell broke loose."

"We fought tooth and nail for the better part of four hours. The NVA fired on us from spider-holes burrowed under root systems of damaged trees. Grenades started falling on us, too, and the automatic weapons fire was so thick it pinned us down. I got hit in my hand and leg, that's when I saw Ted jump from the safety of a trench and assault an enemy emplacement. He wiped it out, singlehandedly.

"We started moving forward again but a rainfall of mortars came down on us. We hit the dirt and covered up, except for Ted. He rushed forward to engage the enemy by himself. That's the last time I saw Ted until we found him the next day. My best friend saved a lot of Marines. I still miss him."

The USMC officially recorded the Siege of Khe Sanh lifting on April 8; one day after Mike MaCauley's 20th birthday.

TED'S MOTHER AND KID BROTHER

Brigadier General Tim Britt; retired: "I was seven years old when my big brother lost his life in Vietnam. He'd pick on me and tease me just like any other big brother, but I always wanted to be where Ted was."

Ted's mother, Joyce Britt: "I had three children. Ted was my oldest son and I lost him in Vietnam, then I lost my only daughter to breast cancer in 2004. Tim's the only child I have left. I sure wish he'd stop riding his motorcycle."

General Britt: "I'm not the only general who saddles a Harley, mom." His thoughts returned to his older brother. "Anyway, it was somewhat rough on my parents when I joined the Army. I guess it would be painful for any parents who had already lost a child in combat. Not a day goes by that I don't think of Ted. He loved to hunt, fish, he loved baseball, and he loved the Marines. Ted enlisted right out of high school."

Joyce Britt: "Ted joined the Marines because he said he wanted to learn from the best. After basic training, he was sent straight into combat at Khe Sanh as a fire team leader. I didn't know what it meant to be a fire team leader, but when I found out it really worried me. But after that, things happened pretty fast. It seemed like he had just got to Vietnam when we were notified that Ted was missing. It took the Marines some time to find my boy before they notified us that he'd been killed. They brought my son home on Mother's Day."

General Britt: "It's like it happened yesterday. I was coming home from school and noticed several cars parked at the house. When I went inside, my sister took me by the hand and led me into the bedroom then told me what happened. I never thought that would happen, not to my big brother."

Mrs. Britt: "I'll never forget that day the Marines arrived at our house. My husband opened the door and immediately realized why they were there. I was upstairs and could hear my husband wailing. I never heard that sound from my husband before but I knew then that Ted…." Mrs. Britt chose not to continue.

General Britt gently touched his mother's arm. "That's okay, Mom," he said, then picked up the conversation. "I remember two Marines at our house, one in particular. He took me under his wing and remained with us until after the funeral. He was like a member of our family, helping dad, running errands; I remember how he kept me occupied during the funeral, trying to take my mind off the situation."

Mrs. Britt: "I appreciate everyone honoring my son."

General Britt: "I have Ted's last letter to our father, if you'd like to read it."

"In peace, sons bury their fathers; in war, fathers bury their sons."
—Herodotus

THE LAST LETTER

January 30, 1968
Hi Dad,

I got your letter today. I was hoping like hell ya'll wouldn't find out what was going on here but I don't guess I can stop the newspapers and news reels. Since you already know about the trouble here at Khe Sanh and you asked me to tell you about it, I'll try to tell you everything that has happened so far.

All the trouble started on the 21st of January around 0500. On the 21st Khe Sanh was hit heavily by mortars, rockets, and artillery. They really did a damn good job on us. They hit our main ammo dump first, causing our own artillery to go off, which blew up half of Khe Sanh. They also destroyed our Headquarters, airstrip, both mess halls, and all our supply tents and bunkers. They really put the hurt to us. Later in the day they shot down two large planes, one Phantom, three helicopters and one spotter plane.

About 0800, a small force of NVA tried to fight their way into Khe Sanh by using tear gas and flame throwers, but we pushed them back and chased them into the hills where we made contact with a whole battalion of NVA between Hill 881 & Hill 861. We called in artillery and air strikes to help us out. Then we got the word there was four regiments of NVA moving towards us so we pulled out and our planes went in and dropped napalm on them. We have been having contact with the NVA ever since the 21st—and it's the 30th now.

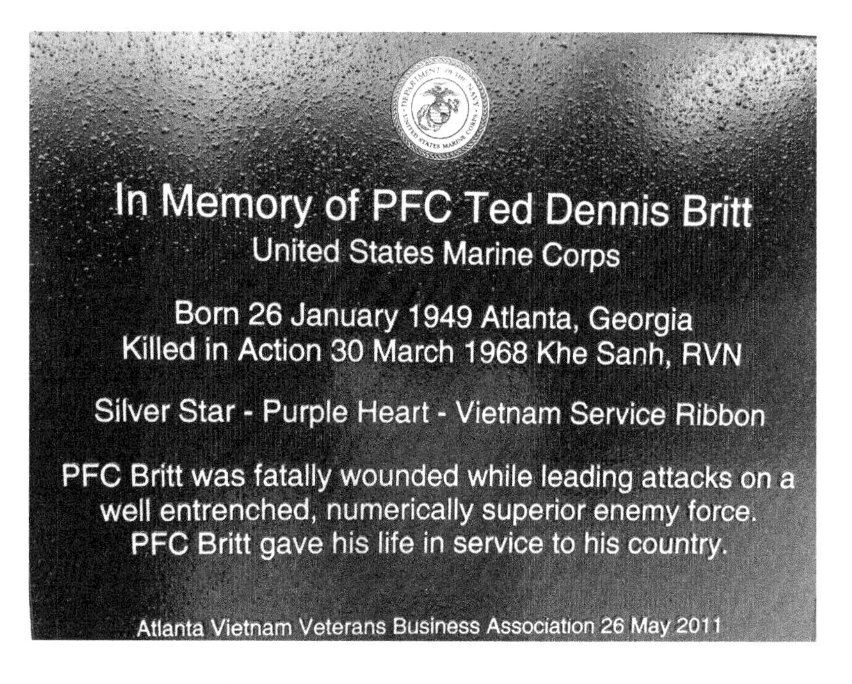

Britt memorial at Walk of Heroes Veterans Memorial

They have been throwing mortars and rockets at us every day since the 21st. They have tried everything you can think of to get inside our lines here at Khe Sanh. A few days ago, they tried using dogs to get inside our barbed wire. They tied satchel charges (TNT) with time detonators on the dogs which were trained to go up to our wire and lay down. The satchel charges would go off, blowing holes in our barbed wire. They also tried tunneling under our wire.

The reason we weren't getting ammo and supplies was because the airstrip was blown up during the mortar attack. The airstrip has been repaired and we're getting ammo and supplies regular. The casualty reports in the papers are correct, the three killed and 39 wounded here at Khe Sanh were caused by the mortar attack on the 21st. There has been one killed and about 14 wounded here at Khe Sanh since the 21st. I can't remember who it was that said, "war is hell" but he was sure right about that. It's an awful and bloody thing to see your friends get blown away right next to you. I just pray that someday the world can live in peace and stop killing each other off like a bunch of savages. I don't think there's a person here on either side that enjoys killing or wants to be killed. I know I don't, but that's what I'm here for and that's my job and I must do my job and defend my country even if I don't think it is right.

Things are starting to cool off a bit now, so there is no need for you to worry about me. I can take care of myself. Besides, I have the deepest foxhole and strongest bunker in Khe Sanh (Ha-ha). Well, I guess I had better close for now so I can get this letter mailed right away. Be good and write soon, and please don't worry about me.

Your loving son, Ted

"Greater love hath no man than this; that a man lay down his life for his friends."

—John 15:13

THE UNSTOPPABLE JOURNALIST

"Their life consisted wholly and solely of war, for they were and always had been front-line infantrymen. They survived because the fates were so kind to them, certainly—but also because they had become hard and immensely wise in animal-like ways of self-preservation."

—Pulitzer Prize WWII Journalist Ernie Pyle

The common soldier loved Ernie Pyle. He told their stories in a down-to-earth style that made the foot soldier feel like someone cared, and they recognized that 'Ernie', as the soldiers referred to the writer, was a war correspondent who appreciated and understood what the label 'infantryman' really implied.

Three weeks before the Japanese savaged the 7th Fleet at Pearl Harbor, a male child came into this world in Bryan, TX, a couple miles north of College Station where his father was attending Texas A&M. He grew up in a small town called Refugio, about 30 miles north of Corpus Christi. Bitten by the journalism bug at a young age, the boy idolized Ernie Pyle. "I wanted to be just like Ernie," he stated. His dream became a violent reality near the Chu Pong Hills at a place called the Ia Drang Valley in the Central

Highlands of Vietnam. In the first major engagement pitting the North Vietnamese Army against American troopers, Joseph Lee "Joe" Galloway flew into the Ia Drang Valley and history, earning him recognition as 'Vietnam's Ernie Pyle'. And this is his story.

Mid-November, 1965, the Ia Drang Valley: The lead F-100 Super Sabre released two napalm canisters but was off-target. The deadly canisters floated idly toward the American lines as if in slow-motion. If the second F-100 released his two canisters in the same manner, the troopers and battalion Headquarters hunkered down behind a large termite mound would be charred beyond recognition. Their Commander, Lt. Colonel Hal Moore, spotted the horror unfolding. He hollered at the top of his lungs at Charles Hastings, the US Air Force ground Forward Air Controller, *"Call that S.O.B off! CALL HIM OFF!"*

Hastings yelled into his radio, *"Pull up! Pull up!"* The second F-100 pulled up just in time, but the first F-100's canisters continued their leisurely tumble straight toward PFC Jimmy Nakayama and Spec 5 James Clark. Seconds later the jellied gasoline ignited both soldiers.

Excerpt from "We Were Soldiers Once…And Young" by Lt. Gen. Harold G. Moore and award-winning war correspondent, Joseph L. Galloway: Wrote Galloway, "Nakayama was all black and Clark was all burned and bleeding. Before that, I had walked over and talked to the engineering men in their little foxholes. Now those same men were dancing in the fire. Their hair burned off in an instant. Their clothes were incinerated. One was a mass of blisters; the other soldier not quite so bad, but he had breathed fire into his lungs. Somebody yelled at me to grab the feet of one of the charred soldiers. When I got them, the boots crumbled and the flesh came off and I could feel the bare bones of his ankles in the palms of my hands. We carried him to the aid station. I can still hear their screams."

One medic lost his life providing aid as other medics pumped morphine into PFC Nakayama and Spec 5 Clark; it didn't stop the pain. Clark survived; Nakayama died two days later, on the same day his wife gave birth to their first child. And the battle for the Ia Drang Valley had just begun.

To understand from an early age your true purpose on God's green earth is a godsend. Joe Galloway is one of those fortunate few; he was one of those lucky kids. Galloway grew up idolizing WWII Pulitzer Prize journalist, Ernie Pyle.

As an aspiring young journalist, Galloway could not have chosen a better war correspondent to revere than Ernie Pyle. His awe for the most respected journalist in WWII would serve him well during his triumphant journey as a war correspondent. Galloway wouldn't admit it, nor accept the accolade, but Vietnam veterans consider Joe Galloway the 'Ernie Pyle' of our generation.

An interview with a legend: I received an unexpected email from Colonel Mark Franklin, History and Legacy Chief for the Vietnam War's Commemoration. Colonel Franklin invited me to sign up for an interview at the Atlanta History Center to relate my story of service in Vietnam. Nada, was my first reaction, until I read the email further to discover the man conducting the interview would be none other than Joe Galloway.

Joe Galloway

Having admired Galloway's writing genius and valor ever since reading his best seller, *We Were Soldiers Once...And Young,* no way, Jose, was I going to turn down an opportunity to meet the living legend. Of course, the real reason yours truly agreed to an interview was to beg, con, sweet-talk...hell, throw a public hissy fit if need be, to take advantage of this once in a lifetime chance to obtain my own interview with Joe Galloway.

As we shook hands before my interview, I noticed that Galloway and I shared the same height and about the same weight. Yet, having gained no height since Vietnam, we both sure as hell had put on a few extra pounds since the war. Yes, we were soldiers once, and incredibly young, but unlike yours truly, Galloway still maintained a beautiful crop of hair. I felt absolutely at ease with the man, as if we had shared the same bunkers, maybe we did. He fired questions in my direction for almost an hour before we switched interview hats. I had Galloway's attention for twenty minutes before his next meeting, too short a time to interview a war correspondent whose life meets all the requirements for a PBS mini-series. Yet his limited remarks spoke volumes on the price of freedom, centering on the siege of Plei Me Special Forces camp which was the precursor to the Ia Drang Valley.

When asked how many wars he covered as a war correspondent, Galloway's reply was a modest yet honest, "I don't even know." A concise synopsis: Shortly after the first US Marines landed at China Beach, Galloway began his sixteen-month tour in Vietnam, starting in April of 1965. He completed two additional tours, in 1971 and 1973, plus squeezed in coverage of the 1971 Pakistan–India War. Fittingly, Galloway returned to Southeast Asia in 1975 to report the fall of Cambodia and South Vietnam. Continuing his splendid career, Galloway reported on six other regional conflicts plus paired up with the Army's 24th Mechanized Infantry Division to participate in their tank charge across the desert of western Iraq during the Gulf War.

In his own words: "I was 23 years old when I began my first of four tours in Vietnam, 1965 to 66, 1971, 1973, plus I was there for the fall of Saigon in 1975. By that time, I hadn't been home in over sixteen years. My assignments included news bureaus in Tokyo and Saigon, chief of bureau in Jakarta, New Delhi, Singapore, Moscow when the Soviet Union was still intact, and finally Los Angeles. Los Angeles is basically a foreign country, too.

"I liked WWII cartoonist Bill Mauldin but totally idolized WWII correspondent Ernie Pyle. I've visited his grave at the Punchbowl in Honolulu. A year ago, I received a newly discovered photo of Ernie that was taken right after he was killed on Ie Shima. Ernie was lying there with probably the first look of peace on his face in four years of war. It just broke my heart." Galloway wept.

"Ernie Pyle was my inspiration, and if my generation had to go to war then I was going to cover it. Perhaps in doing so it is easier 50 years later to explain why you went to war instead of why you stayed at home. At the time, I was working for UPI in Topeka, Kansas, covering the legislature, state politics, and murder trials. The year was 1963. I wrote my bosses every week demanding, pleading, begging, for a transfer to Southeast Asia, which meant Vietnam. We were going to have a war there, my generation's war, and I wanted to cover it. Frankly, I raised holy hell with my bosses.

"Then after the 1964 election, Lyndon Johnson promised that American boys would never be sent to do for Vietnamese boys what they ought to be doing for their own country. Well, I knew Lyndon Johnson because I came from south Texas and I knew he was lying...we were going to have a war and I was going to be there. UPI eventually called and transferred me to Tokyo. About four months later, I landed in Saigon right after the 1st Battalion, 9th Marines landed in Da Nang. I spent two days in Saigon, got a Press Pass, then hooked up with the Marines for seven months."

October of 1965, the Plei Me Special Forces Camp in the Central Highlands, Galloway joins the fight: "The North Vietnamese who had come down the Ho Chi Minh Trail in Laos, the same ones we eventually fought in the Ia Drang Valley, had surrounded the Plei Me Special Forces Camp to bait the South Vietnamese Army, the ARVN, to come down the road from Pleiku as a relief column. Their idea was to ambush the ARVN's relief column, then overrun the camp. A regiment of the bastards was out there surrounding the camp and another regiment was lying in ambush.

"I wanted to get in there but the air space was closed. They'd already lost two Hueys, two Air Force B-57 Canberra bombers, and an A1E Skyraider. I'm stomping around the flight line at Pleiku trying to find a ride when I met up with my buddy, an old Texas boy, Capt. Ray Burnes, from Ganado, TX. He said, 'What's the matter, Joe, you look upset.' I told him I was looking for a ride into the Plei Me camp. Ray replied, 'Hell, Joe, the air space is closed. Man, they ain't letting no flights in there.' I told him I didn't care, I wanted to go anyway. He said, 'Well, I'd like a look at it myself. Com'on, I'll give you a ride in my bird.' So we flew to Plei Me.

"I have a photo taken from the chopper as it laid on its side to spiral in. The camp was triangular. You could see exploding mortar rounds and I thought, 'Geezzzz, is that where we're going?' As soon as we landed, I jumped off the Huey. They started tossing in wounded Montagnards then Ray lifts off, smiling at me, and shooting me the, well, you know, his longest finger. A Special Forces master sergeant ran up to me and said, 'Sir, I don't know who the hell you are, but Major Beckwith wants to see you right away.' I asked who that was, and he answered, 'that big fellow over there jumping up and down on his hat!' And I was thinking, 'this ain't good.'

"So I went over to Beckwith who immediately demanded,

'Who the hell are you?' I replied that I was a reporter. Beckwith was furious, 'I need everything in the damn world, medevac, food, ammo, and what has the Army sent me in its infinite wisdom…a God-forsaken reporter. I have news for you, son, I have no vacancies for a reporter, but I'm in desperate need of a corner machine gunner, and you are it!' And I thought, 'this ain't too good either.' I was given 10 minutes of instruction on how to care for and feed a .30 caliber machine gun. I stayed on the wall for three days and two nights until the relief column finally fought its way into the camp." Note: then-Major Charles Beckwith would best be remembered as the creator of the premier counter terrorism unit, the Delta Force.

Galloway continued, "The 7th Cav. at long last relieved us. I recall Beckwith asking me, 'you did a good job on that machine gun, son, how'd you like to come with me on a recon mission?' I asked him where to; Beckwith pointed west and stated, 'Over yonder.' Over yonder was Cambodia, so I replied, 'how long will we stay?' and he replied, 'Oh, a couple of weeks.' I thought a moment then said, 'Major, I'd love to but I'll probably be fired for being stuck in this camp for three days and out of touch with my office in Saigon.' Put it this way, I didn't go into Cambodia." Galloway hooked up with the 7th Cav.

For the next few days then Lt Col Hal Moore and his 7th Cav. troopers, with renowned Sergeant Major Basil Plumley barking instructions, ran an operation in the hills east of Plei Me. Galloway: "That was pretty much a walk in the sun, or better yet, a slow crawl in the jungle. We humped all day, forded a mountain river right before dark, got soaking wet, made camp and dug a hole, wrapped ourselves in our ponchos and almost froze to death. I have never been colder than at 4,000 feet in those Mountains in Nam.

"The next morning I'm hankering for some hot coffee, so I'm in my little foxhole cooking up a can of water to pour dry coffee into it, but I glance up and see boots at the edge of my hole, four of them. Those boots belonged to Lt. Colonel Hal Moore and Sergeant Major Plumley, another of those, 'this ain't good,' situations. Moore said, 'Son, I have to tell you, in my battalion we all shave in the morning, and that includes odd reporters.' I looked down at my hot water, Moore's gawking at it....so I repurposed that canteen cup of hot water. That was my first face-to-face meeting with Hal Moore."

Sunday, November 14th, in the Ia Drang Valley: Galloway: "It's the first day, first lift of the big battle. I had my first look at LZ X-ray (Ia Drang base camp) an hour after the 7th Cav. landed. I was riding with their Brigade Commander, Colonel Brown. LZ X-ray was easy to spot; the smoke was 5,000 feet in the air. We circled the vicinity while Brown and Moore argued. Brown wanted to land, but Moore kept saying if we did land in the command chopper with all those antennas sticking out we'd be walking home, that the NVA would shoot us up.

"While they're trading words, an Air Force A1E Skyraider, engulfed in flames, passes under our chopper and crashes into the jungle. Everybody yells, 'Did you see a chute, did you see a chute?' It took place on my side of the chopper and I'm on the net saying, 'No chute, no chute, he went in with the plane.' I later found out his name; Captain Paul McClellan, Jr. I know where he is. The MIA/POW organization called his wife about 15 years ago. They said they could get to the crash site to do a dig, maybe uncover his remains. She said 'no', that her husband loved what he did and died in a battle that has gone down in history, and to let him rest in peace. She indicated her four kids were helped through college by the US Air Force. She was at peace, too."

Joe Galloway's final thoughts, final comments: "The survivors of the Ia Drang Valley, the men of the 1st Battalion 7th Cav., will be my friends for life. I saw them fight and die beside each other and for each other, I heard the command 'fix bayonets' and saw those bayonets used on human beings. I had men killed on my left and right, and I figured every man who died there, every man who was wounded there, were casualties that saved my life."

The Hollywood version of the Ia Drang battle depicts Galloway being handed an M-16 while being advised, 'prepare to defend yourself.' He stated, smiling, "That's not accurate. I carried one in, I 'brung' my own."

American losses at LZ X-ray and surrounding support bases, 308. NVA losses, around 1,700 soldiers and an unknown number of wounded. In the battle, American artillery alone fired more than 18,000 rounds in 53 straight hours of support for Hal Moore and his courageous troopers.

From the book, "We Were Soldiers Once…And Young"—page 200. 'As they left the Ia Drang Valley, Moore and Galloway came face to face.' "We stood and looked at each other and suddenly without shame the tears were cutting tracks through the dirt on our faces. Moore choked out these words, 'Joe, go tell America what these brave men did; tell them how their sons died.'"

Joe Galloway filmed and fought and penned the account of the Ia Drang Valley. Impressed by his deeds, UPI raised Galloway's salary from $135 to $150 a week.

On a personal note: I had the honor to interview a legend, the man called Galloway, but behind the legend I discovered a real human being, quick to joke, yet quicker to shed an open tear recalling the memories of the heroes that still haunt his soul.

God be with you, Joe… a job well done, sir.

THREE DOWN AND ONE TO GO

"Above all, Vietnam was a war that asked everything of a few and nothing of most in America."

—Myra McPherson

After a gentleman from our church began a small veterans group, it was obvious to every veteran in attendance that we were in need of informed leadership. We fell short in goal and direction, plus several of the guys asked about benefits or problem solving solutions that none in attendance could offer from experience. Grumbles and gripes and genuine bullshit pretty much dominated the meeting. Frustrated, I suggested that we invite Tommy Clack to our next meeting to speak with the group. I'd never met the gentleman, but certainly knew of his reputation.

Scuttlebutt had it if a veteran from our county or the six neighboring counties needed well-informed assistance with VA benefits and/or problems, the place to visit was the local office for the Georgia Department of Veterans Service. And the man to see was Tommy Clack.

Meeting Tommy for the first time requires emotional control, restraint from making a dim-witted comment, and an ability to overcome the awkward moment when his left hand is offered for a handshake. Having lost both legs plus his right shoulder and arm in Vietnam, the appearance of this one-armed workhorse for veterans has a tendency to arouse pity, but pity the individual that pities Tommy Clack.

Vehemently independent, a life-long confinement to a motorized wheelchair has not dampened his work ethic nor his love for hunting and fishing. Lesser men do lesser things; Tommy Clack is a better man. Flawlessly dressed, debonair good-looks enhanced by a neatly trimmed beard, the diction of a Rhodes Scholar, and an all-business approach on the job camouflages the personality of a witty fun-loving guy. On dating women: "I may not be the man I once was, but I can be the man I was at least once; twice if she gives me enough time."

He agreed to speak to our church veterans group and did so with the clout, confidence, and courage of a true American patriot with zero tolerance for political correctness, which Tommy considers nothing more than sneaky old-fashioned political control. Brutally blunt with truth and fairness, Tommy is of the rare breed once cherished and respected in America before fashion and foolishness and politics turned men into mice. I made it my goal to make this man my friend.

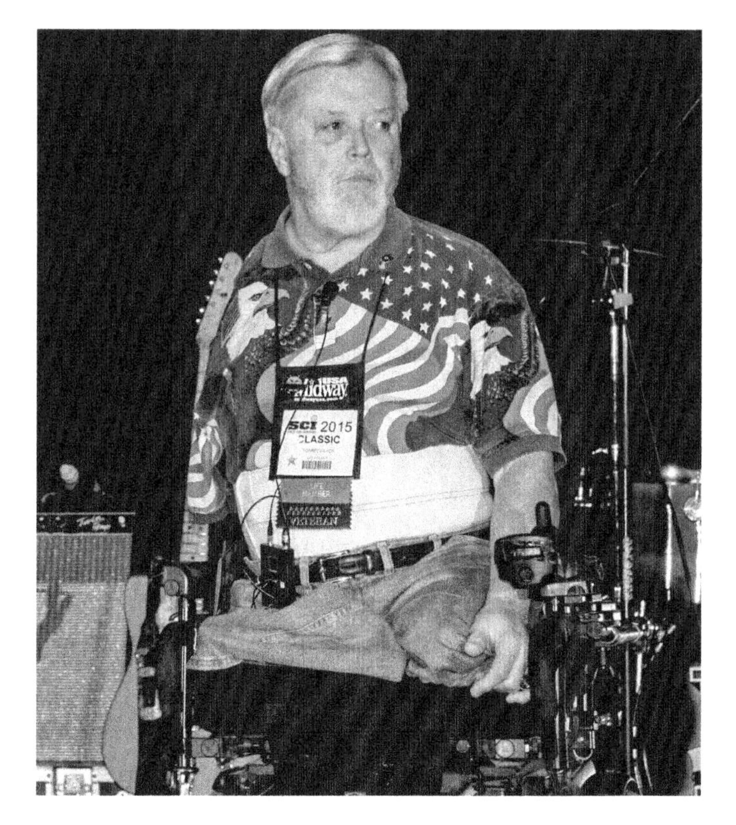

Tommy Clack, an American Hero

I accomplished my goal and ever since, he's worked my ass off on several veteran projects. That characterizes Tommy: veterans, first and foremost. The man even conned me into being master of ceremonies at a 4th of July patriotic event. His con-artistry knows no bounds.

Tommy asked me if I would consider penning a book about his life. I had envisioned the project many times. His story needs to be told, and it needs to be remembered. The title I suggested: *Captain Clack*. The Captain; however, had his own inspiration. "Nope," he replied. "I already know what title I want. Let's call it, *Three Down and One to Go.*"

Ya gotta love this guy.

CAPTAIN CLACK

A graphic on the front of his business card depicts a bald eagle in flight; beneath the soaring eagle a reference from the Holy Bible: John 15:13 — *Greater love has no one than this, than to lay down one's life for his friends.* At the top of the card four words reflect his sense of humor: Have Wheelchair Will Travel. And words in unpretentious print on the back top-half authenticates his credentials:

Captain U.S. Army, Retired
Combat Wounded / Vietnam / May 29, 1969
Forward Observer, C/2/27 Inf., 25th Inf. Div

A phrase on the bottom-half caught my attention. Tommy explained, "That's the definition of my attitude that I've carried through my life. I coined the phrase in the early 70's." Planet Earth would be a much better place if we could all espouse to his coined phrase.

THE PHRASE

Our lives are not determined by what happens to us but by how we re-

act to what happens; not by what life brings to us but by the ***attitude*** *we bring to life. A positive* ***attitude*** *causes a chain reaction of positive thoughts, events, and outcomes. It is a catalyst, the spark that creates extraordinary results.*

Extraordinary results spawned by an extraordinary man and combat veteran. A recipient of the Bronze Star, Purple Heart, two Air Medals, and numerous Vietnamese medals, thus far in civilian life he's been the recipient of over two hundred prestigious local, state, and federal decorations and honors. He's covered all fifty states as a public speaker with over 7,000 speaking engagements, been one of the subjects in five dissimilar books, and one of the subjects of the TV series *In Search Of* hosted by Leonard Nimoy of Doctor Spock fame, dealing with Near Death Experiences. He even found the time to own and sponsor a 1967 Camaro D-gas drag racer for four years, painted with a red, white, and blue color scheme, of course. Yet for all the decorations and tributes, he considers his greatest accomplishment was being a single parent and raising two children, his son, Adam, and his daughter, Erin.

Johnny T. (Tommy) Clack was born and raised in Decatur, Georgia. His father, a decorated veteran of World War II and the Korean War, was just one of many in a family steeped in military tradition. In 1961, when the first American advisor lost his life in the Republic of South Vietnam, his death inspired Tommy to write a term paper for his 8th grade history class entitled, *This is My War*. His destiny lay before him. A 1965 honor graduate of Decatur High School, his recognition as Most Outstanding in Track and Field earned him a track scholarship to the University of Houston.

Albeit aware that the intensity of the conflict in Southeast Asia was mounting daily, 8th generation Army Johnny T. Clack dropped out of college during his freshman year and volunteered for the United States Army. His entire group of enlistees was sent to Officer Candidate School. One 2nd Lt. in his graduation class earned the distinction as "Most Likely to Make Four Star." America would come to know Tommy's classmate as Four Star General Tommy Franks.

Tommy requested, and received, training in Army artillery. Smiling, he admitted, "As a kid I always enjoyed shooting off firecrackers so I figured artillery offered me more bang for my dollar." He excelled at his new assignment, so well that his petition for frontline combat duty in Vietnam was denied. Apparently, Uncle Sam needed officers infatuated with explosives to train other firecracker aficionados: 2nd Lt. Clack was assigned as an artillery instructor.

He recalled, "I trained artillerymen from all over the world, including Israeli soldiers after the 1967 Middle East War. But in time, I convinced my superiors that I'd be a lot more proficient as an instructor if I had combat experience." Tommy's feisty tactics and cogent doggedness for combat gained him his first tour of Vietnam as a FO (forward observer) with the Vietnam Field Force. "I covered about every base camp from the DMZ to the Mekong Delta," he said. "Plus, I trained American, South Korean, Australian, and South Vietnamese artillerymen on use of the newest artillery round, a CBU or Cluster Bomb Unit. The CBUs were packed with submunitions that were released on impact."

Promoted to Captain, Tommy was serving his second tour of Nam with the C/2/27 Wolfhounds of the 25th Infantry Division when tragedy struck on May 29, 1969. He recalled, "We were engaged in a hellacious firefight on the Cambodian border near the village of Ap Duc, southwest of the Nua Ba Dien (The Black Virgin Mountains). We had taken up a blocking position on the main road, which is Highway One from Saigon to Phnom Penh, when they tried to overrun us. But we hunkered-down and denied them passage. I was in the process of calling in artillery support when an RPG hit my right foot and detonated."

An RPG (rocket-propelled grenade) has an effective range of 150 meters (500 feet) and deploys an 82mm warhead. Tommy continued, "I was awake and alert for about five minutes. I knew I was missing three limbs because I could see them scattered about me, but due to the continuous loss of blood, my vision and awareness faded until I couldn't hear nor feel the people trying to work on me." At a recent Wolfhound reunion, one of the men who tried to stabilize Tommy that day, Staff Sergeant Quintrell, confessed that after they loaded Tommy onto the medevac chopper they all prayed he would go ahead and die due to the extensive damage they saw on his body.

Captain Tommy Clack had lost both legs above the knee, all his right shoulder and arm, suffered massive internal injuries, a loss of hearing, and his right eye dangled like a Christmas ornament on his cheek. Unable to find a pulse or any signs of life, the medic aboard the chopper covered Tommy's lifeless body with a rain poncho. Upon landing at the 12th Evac Hospital (MASH Unit),

Tommy was taken off the chopper and laid in the line of other dead soldiers. The only good thing the RPG did that day was to cauterize his horrendous wounds with the force of the explosion.

Tommy said, "After the doctor had saved those he could in the hospital trauma room, he came out and walked over to the line of dead bodies covered with ponchos. The doctor does not know to this day why he lifted the poncho over my body, but he did. He saw something that made him realize that I was still alive. They rushed me inside to the trauma room and, well, here I am today. I've talked to that doctor several times and he and I both agree that God led him to my body lying there."

Thus began a 22 month hospitalization, 20 of those months at the Atlanta VA Medical Center. Tommy was placed on D Ward, the D meaning Death. From May of 1969 until October of 1969, Tommy survived on life-support. From a 6'2" and 190 lb. physique, the cost of war had dwindled his body to 55 lbs. of flesh and bone. He was not expected to live.

But God had other plans for Captain Tommy Clack. The tubes came out in October; he learned to sit up, he learned to crawl, he learned his center of gravity, and sat in his first wheelchair in January of 1970. Improving steadily, he was visited by future senator Max Cleland. Tommy explained, "Max is also a triple amputee from Vietnam. He gave me the inspiration to move on with my life and to live it to the fullest."

After 33 operations, Tommy walked out of the VA Medical Center on March 21, 1971. "That's right, I 'walked' out," Tommy said. "I was walking on artificial legs weighing 20 lbs. apiece, but by golly, I did 'walk' out of that place." He paused briefly, then continued, "I guess if there is anything 'good' developing from our present conflicts, especially in Afghanistan and the Middle East, it's the speed of medical assistance for our wounded. They receive treatment almost immediately in combat; can be in Germany within twelve hours, and arrive home in America within thirty-six hours."

Today our wounded warriors walk on prostheses weighing about 3 pounds. But for Tommy, with only three inches of femur remaining in his right leg, it was impossible in 1971 to adapt to artificial legs. A wheelchair would provide the mobility he needed for the rest of his life. Since leaving the VA Medical Center, Tommy has spent three years of recovery time from an additional 17 inpatient surgeries and 15 outpatient surgeries. Yet, even as he endured prolonged periods of adjustment, routine pain without knowing when his next surgery would be scheduled, Tommy still found the time and internal fortitude to complete his education. In 1975, Captain Johnny T. Clack graduated with honors from Georgia State University with a Bachelor of Science Degree in Urban Administration and City Management.

"God gave me a second chance," he said. "Personally, I believe I'm a better person for what I endured. I'm a better person with one appendage than I was with four."

In the 1970's, Tommy worked with Georgia Senator Sam Nunn and members of Congress on veterans' affairs. "I found my niche in helping other people," he said. From February of '79 to May of '93 Tommy was employed as Staff Assistant to the Director of the Atlanta VA Medical Center, plus served as the president of a volunteer in-house civic group, The Employees Association. In 1983, Tommy served as Director of the federal sector Combined Federal Campaign within the metro Atlanta United Way Fund Drive, raising $1.2 million dollars.

Since June of '93, he has served as a Field Office Manager at the Conyers location for the Georgia Department of Veterans Service. "Eighty to a hundred work hours per week is the norm around here," he said. "This year our contacts will be well over 31,000. This is what I do. My life is working with veterans, their widows, and families. I love my job. If I kick the bucket right here at my desk, then that will be okay with me."

Tommy steadfastly believes all veterans have three tasks:

1. Perpetuate our military legacy as veterans; to keep the truth flowing in schools and civic groups.
2. Maintain the relentless search to account for and/or recover all American warriors M.I.A. (Missing in Action). SPECIAL NOTE: Of the original 1,973 MIAs at the end of the war, 1,256 remain MIA in Vietnam. In Laos: originally 573—295 remain MIA. Cambodia: originally 90—48 remain MIA. China (territorial waters)—originally 10—7 remain MIA.
3. Raise our children properly to instill personal responsibility, patriotism, and accountability so our future generations live and work in a civil society.

"I'm a strong believer in competition," Tommy said. "Only by failing can we better ourselves. There are winners and losers, that's life, that's how the world operates, and we're doing our children a big disservice by letting every kid win and nobody loses."

One would believe Tommy as the embodiment of the word "hero", but he stalwartly rejects the suggestion. "I'm not a hero and frankly I don't know of any heroes except for the men and women that gave their all. All their names, over 58,300 of them, are on a long black wall in Washington, D.C. Those are my heroes."

The technology of war has improved: smart bombs, cruise missiles, unpiloted drones, night vision, stealth aircraft, an unbelievable array of killing machines, but the sacrifice and length of combat exposure has not. Tommy explained, "In World War Two, our soldiers averaged 40 days of combat exposure, in Korea 180 days, in Vietnam 240 days, and in the Iraq War and now in Afghanistan 310 days. It is true, freedom is not free." During his hospitalization at the VA in March of 1970, Tommy was credited for originating the slogan "Freedom isn't Free." The phrase is now honored nationwide.

If a veteran has an issue or misunderstanding with the VA, Tommy Clack is The Man. He maintains, "The VA is in the business of saying 'no' and to minimize, but I'm in the business of finding a way for the VA to say 'yes' and to maximize." Tommy is all-business when working for and in defense of veterans, even to the point that some people consider him 'cold' or 'unfriendly' and certainly lacking a sense of humor. Do not believe one word of it!

He tells the tale: "I spent the 1970s with the Jaycees and during Halloween we would set up a haunted house at the Belvedere Shopping Center in DeKalb County. I would always be positioned in the last spook room. It would be pitch dark; we had rotten meat lying around to make the place smell really sickening, then suddenly strobe lights would come on and I'd be lying there acting injured. All of a sudden, this scary looking dude would come out with a chainsaw and cut off my artificial legs, it was the Jaycees' version of The Texas Chainsaw Massacre. Blood, well I mean ketchup, would be all over the place. People would scream their heads off, absolutely horrified. I thought it was hilarious."

Another entertaining episode (his words, not mine) was on Jekyll Island in 1975 when the movie *Jaws* was terrorizing audiences and sending beach real estate into an economic tailspin. Again with the Jaycees, Tommy and his partners in crime secured him into a lounge chair and surreptitiously slipped Tommy into the ocean then at the proper depth they set him down, waded out, and left Tommy alone, neck high in the frothy breakers. After waiting patiently for an ample quantity of naïve sun-worshipers to come within hearing range, Tommy punctured balloons filled with red dye then started screaming, "SHARK! SHARK!"

The Jaycees ran to the rescue! They dragged his crimson-covered body from the surf with, observably, three missing limbs. Tommy said with a mischievous grin, "The beach bums freaked-out and one lady even barfed on the sand, but when they found out it was a practical joke, somebody was so pissed off they called the police. Well, two cops showed up but they didn't know what to do to me. I finally asked them, 'Well, am I under arrest or what?' One officer shook his head and replied, 'No, sir, we don't really know what to charge you with. But if you do this again, we are going to arrest your ass for something!' I thought it was hilarious as hell."

Tommy has played movie roles depicting wounded soldiers with missing limbs. He also taught himself to swim again. "I'd put on my trunks, get to the edge of the pool, and jump in. I figured I'd either sink or swim," he said. "Lifeguards would try to save me because they thought I was drowning, but of course if the lifeguard was an attractive female, well, I'd let her. Mouth to mouth resuscitation was okay, too." At a few public pools, Tommy was asked to leave. Parents considered his appearance unacceptable for their children. Tommy explained, "The parents would leave and complain to the pool management, but when they came to throw me out, their children were already gathered around me to ask questions. The children didn't have any problem with me, their parents did."

Amputations and life-threatening injuries from explosive devices in Vietnam were 300% higher than in WWII; 70% higher than in Korea. Of the injured, 5,283 were amputees, 1,081 were multiple amputees and 73 were triple amputees, yet had it not been for the medevacs and quick availability to surgical hospitals, the fatality rate would have been much higher. A wounded soldier in Vietnam had a much greater chance of survival than his counterparts in Korea and WWII.

The motto of the United States Army is *haec protegimus* (This We'll Defend). It certainly applies to a man the fog of war once pronounced dead, yet he fought for life and has used his new life to defend his fellow veterans, to support the cause of freedom, to honor his country and his fellow countrymen, to take the high ground, to never capitulate.

In the touchy and complicated arena of veterans' issues and an on-again off-again military readiness dependent on what political party is in power, I am honored to stand beside this man at any-place, anytime, anywhere.

This We'll Defend, Captain.

Special Note: Since this writing, Captain Tommy Clack has re-tired from the Georgia Department of Veterans Service to devote his energy and time as the main driving force for the completion of The Walk of Heroes Veterans War Memorial in Rockdale County, Georgia. A moving and magnificent memorial, the 35-million dol-lar project will be unlike any other War Memorial in the United States. Please visit their website: www.walkofheroes.org

CARIBOU TO SANTABOU

Born and raised in Cajun country, Alan Gravel received a de-gree in Civil Engineering from Louisiana Tech before obtain-ing a Master's Degree in Environmental Health Engineering at the University of Texas. February of 1969 launched 23-year-old Gravel into officer training on the Medina Campus at Lackland AFB in San Antonio, Texas. He recalled, "I wanted to fly, so the Air Force sent me to Laughlin, AFB in Del Rio, Texas to see if I could qualify. The first aircraft a trainee had to master was the T-41 Mescalero, a military version of the Cessna 172."

Once qualified on the Mescalero, Gravel sat behind the controls of the T-37 Tweet Jet trainer before moving up to the supersonic T-38 Talon. "The training took a year," he said. "Then I was sent to Abilene, Texas for training on the DeHavilland Canada C-7 Caribou. The Caribou was a STOL aircraft, Short Takeoff and Landing. I attended survival school in Spokane, advanced survival school in the Philippines, after which I flew into Cam Ranh Bay, Vietnam in September of 1970."

Deemed an orphaned child of aviation, the reliable Caribou did not receive the proper recognition it truly earned in Vietnam. First purchased by the Army, in 1967 the Air Force took over the Caribous while transferring most of its choppers to the Army. The Air Force had utilized Caribous for two years by the time Gravel set foot in Vietnam. He recalled, "I joined the 536th Squadron. I flew copilot before earning the pilot's seat. We'd fly into Bien Hoa for a seven-day stage, meaning we'd fly out of there for three days, get one day off, two more days of flying, then fly back to Cam Ranh. We also flew a three-day stage mission out of Can Tho, but without days off. The Caribou served 75 airfields where no other fixed-winged cargo airplane could go."

Recalling his C-7 training in Abilene, Gravel said, "Our instructors told us, 'You will be shot at, land on awful airfields, deal regularly with tough situations like finding a live hand grenade on your aircraft, but it will be the weather that kills you.' Well, they weren't pulling our leg. To do our job in Vietnam we had to do things on the dangerous side, like flying too low through dense clouds or thick fog, and if on instruments there was the danger of smacking into the side of a mountain. Then you had to factor in the so-called airfields. A Caribou needs about 1200 feet of runway, and that's short, but one Special Forces airfield was 900 feet long and 40 feet wide. If you're going in with a full load, that's pretty tight."

Gravel's air drone directory described the crude Dak Pek airfield as, *'A Caribou's wing will clear the hill as long as your wheels are centered on the asphalt.'* Gravel said, "That meant stay in the center or else." The airfield at Dak Seang was nicknamed 'the ski slope'. Gravel explained, "The airstrip resembled a hunchback snake; it had a sizeable dip in the middle, like a small valley. You couldn't see 'down in the 'valley' from either end of the airfield. Some runways had roads running across the middle and some roads were used for runways."

Advised to fly at 3,000 feet to avoid ground fire, Gravel said, "I believe that's the only thing we did by the book. We were taught that if you're coming in to land and things didn't look right, then 'go back around' to try a second time. Not in Nam. To 'go back around' you had to add power, which meant you make a lot of noise, which attracts a lot of attention. You would be basically telling the bad guys, 'Okay, we're going to give you guys enough time to set up your mortars.' No way, Jose. I'd flown with a real cocky but skillful pilot who taught me a terrific tactical approach. You come over the runway as slow as possible, put down your landing gear then spiral down from 3,000 feet. You roll out at 1,000 feet, run up the engines enough to set propeller RPMs for a 'go around' if ever necessary, then back to idle with flaps down. With enough know-how, a Caribou jock didn't even have to touch the throttle, he was basically landing with a dead stick. We'd receive small arms fire, but didn't realize it until we landed and even then someone had to tell us."

The tough little Caribou mastered the crude airfields at Song Be, Du Dop, Thein Ngon, Katum, Loc Ninh, and Djamap. Gravel said, "One pilot landed in the wrong place and didn't have the room to get back out. A Sikorsky S-64 Skycrane was called in to lift him out. That probably embarrassed the Caribou pilot, but not near as embarrassed as the Skycrane crew. After they hooked up the stranded Caribou and got airborne, they dropped it. The aircraft was a total loss."

Caribous were utilized for 'maintenance alerts' at Cam Ranh Bay. "We had a Caribou loaded and ready to go," Gravel said. "When a Caribou broke down way out in the boonies, the maintenance Caribou would fly to them, swap planes so the original crew could complete their mission while the maintenance team and crew waited for the downed Caribou to be repaired. On one maintenance mission to Ban Don, we found a Green Beret advisor who had been waiting three days for extraction back to Nha Trang. We repaired the Caribou then gave the poor guy a hop back to his base. The incident reminded me why I didn't join the Army."

Caribou to Santabou: On Christmas Day, each squadron painted a Santa Claus face on a Caribou. The radome, a round radar apparatus on the nose of a Caribou, was a perfect Santa's nose when painted red. Gravel recalled, "I remember Christmas Day, 1970. We were flying to Djamap when I heard a Santabou pilot calling ground control, 'Djamap, this is Santabou 420 inbound for landing.' Well, that totally bewildered the Army guy on the ground. He finally replied, 'Uh, yeah…okay,' then the radio fell silent. He'd never heard of a Santabou. I radioed in minutes later, 'Djamap, this is Iris 416, inbound for landing.' The Army guy came back immediately, 'Right, Iris 416! Uh, Iris 416, did you just hear something or someone else on this frequency a few minutes ago?' I couldn't lie to the poor guy, so I suggested that he and his buddies get ready to have a good time because a Caribou Santabou was inbound with nice-looking Donut Dollies, cookies, cakes, egg nog… well, talk about an excited Army man! We touched down after the Santabou and noticed that the party had already started. We never, ever shut down at Djamap, but we sure as heck did that day!"

Caribou converted into a Santabou

Surviving his one year tour, Gravel's next port-of-call was Castle AFB in California for flight training on the KC-135 Stratotanker aerial refueling aircraft. In May of 1972, he was back in the Philippines at Clark AFB and back in the war. Gravel: "We'd fly from Clark and set up a 60-mile refueling track called Purple Anchor, just north of Da Nang and along the coast of North Vietnam. We mostly refueled the fast movers (fighters), like the F-4 Phantoms, sometimes individually, sometimes in a flight of four. One tanker pilot I flew with was an ex-fighter pilot and he wasn't real happy flying tankers, but he sure proved his worth."

Gravel described the incident: "A Phantom F-4 fighter, shot up and losing fuel, called in a 'Mayday.' He was in really bad shape, losing fuel so fast we knew he couldn't make it back to Da Nang. We were his only hope so we went after the guy, meaning we went inland, something we were ordered never to do. We painted him on our radar screen heading directly toward us so we did a 180 and waited for him. He pulled up to gulp fuel but the boom operator couldn't connect with his fighter because the F-4's latch mechanisms on the Phantom's receptacle wouldn't open. We attempted a procedure that is dangerous but doable, applying pressure on him as he puts pressure on us, meaning he's nudging his aircraft towards us as we nudge ours towards his, with only the refueling boom keeping us apart. I normally used two of the six pumps to refuel; in this case, I turned on all six. With that amount of fuel rushing out we were able to blow a portion of fuel past the damaged receptacle to at least replace what he was burning to stay airborne. If the boom had snapped from the pressure, we wouldn't have had the time to respond, meaning both aircraft would have collided in midair. Luckily, he was a great pilot and we had a great boom operator. We flew right over Da Nang and dropped him off.

By the time he touched down, the Phantom was out of fuel. We don't have a clue to his name, his unit, where he was based, nothing. It's called teamwork, getting the job done."

Alan Gravel, on his final flight, flying naked

Note: There is at least one other 'known' incident during the Vietnam War when a refueling tanker disregarded strict orders and rules of engagement to assist and/or save another aircraft in distress. Violations of such severity can result in a court-martial, but in both cases mentioned, the incident was simply 'forgotten'.

Gravel at Cam Ranh Bay - 1971

Gravel continued, "On slow days, the fuel tankers flying Purple Anchor would stack up. Active refueling at 16,000 feet, tankers on standby at 20,000 and 24,000, and sometimes even 28,000 feet. Chicks (fighters) would come up to refuel until the bottom tanker hit bingo fuel and was forced to head home. Then we'd rack down 4,000 feet to the next level. Shoot, those slow-day missions could easily turn into 16 hour ordeals." Unfortunately, one KC-135 was lost out of Okinawa during Gravel's second tour. It simply disappeared. Formal pretext: 'The cause unknown; lost over the Pacific.' The copilot was a graduate of North Georgia College.

His duty done, Gravel left the military and moved to Atlanta so he and his wife could be near family. Utilizing his engineering talents, Gravel eventually started his own business, Willow Construction, completing such projects as the reservoirs for Clayton County and Peachtree City, dam building and repair, and water and sewage facilities. The Gravels lost a son to leukemia in 1980.

In conclusion, Gravel stated, "I know it was war, and it may be difficult for people to understand, but I enjoyed my missions, in the Caribous and the tankers. The soldiers in the rice paddies or jungles and the chicks in the air depended on us to do our jobs correctly so they could continue to do theirs. In my opinion, it's those guys who deserve the recognition and respect."

THE CHEMIST

Born in 1929 in Toledo, OH, a short six weeks later, Richard Grimes and his family moved to White Plains, GA. He recalled, "My dad had a large dairy farm in White Plains, several hundred acres. He delivered milk to markets in Atlanta and Augusta. We did well, even though our country was gripped by the Great Depression. However, by 1935 my father was pouring milk down a sewer because the milk market bottomed. Luckily, he found employment with Toledo Scales in Ashville, NC. And, yes, we took a milk cow with us."

As the winds of war swept across Europe and Asia, Toledo Scales home office transferred Grimes' father back to Ohio. "We understood a war was on the horizon," he said. "The German military victories had been in the news for a long time. We even heard the news in church. I was at a Boy Scout camp on December 7, 1941. Our scout master told us about the attack on Pearl Harbor and we were all old enough to know that it meant our country was at war."

Major Grimes in Vietnam

Teenager Richard Grimes kept track of WWII. He said, "It's as vivid as yesterday. I followed the news of Rommel's defeat in North Africa, the D-Day Invasion of Normandy, Iwo Jima and Okinawa, then of course the two atomic bombs being dropped. I was 16 years old at the time and knew when I turned 17 I'd be gone, probably for the Invasion of Japan. If it hadn't been for those two atomic bombs, well, I probably wouldn't be here today."

The family returned to Georgia after the war. Grimes said with a smile, "Dad didn't care too much for Yankees. Anyway, I finished high school in 1947, the Salutatorian of 15 graduates at Arabi High School. I wasn't the smartest, just the least dumbest."

Grimes received a scholarship to the University of Georgia. He recalled, "Back then all males took ROTC. I majored in chemistry and graduated as a distinguished military graduate in 1951." Due to his extraordinary accomplishments, Grimes received a regular commission in the Army instead of a reserve commission. A year earlier, North Korea invaded South Korea; a war was on. "I figured that would be my next port-of-call," Grimes said. "While I was awaiting orders, I found out I could go to jump school at Fort Benning. So I did."

Country boy 2nd Lt. Richard Grimes was soon jumping out of airplanes. "Well, jump pay added $100.00 to my paycheck and as newlyweds we sure needed it." Grimes met his wife, Ann, at the University of Georgia. "Ann was an English major. That's a good fiancé to have if you're going to college," he said, grinning.

On jumping out of planes: "It's not a natural thing to do, that's for sure. But the training was excellent; jump school was my favorite. Your first three seconds after jumping feels like an eternity, but serene, peaceful, very quiet. Then, wham! The chute opens. We wore shoulder padding to prevent bruising. Our padding...well, the padding was store-bought and very feminine."

Fort Knox, KY: "At Fort Knox, I took an 18-week basic officer's course in armored tactics and recon. We learned map reading plus pioneering. We utilized two tanks, our main battle tank the M4E8 Sherman with a 500hp engine, and the smaller M-24 General Sheridan with two Cadillac engines. It's like comparing a Model-T to a Corvette."

Fort Campbell, KY: "I joined the 11th Airborne, 710th Tank Battalion as a platoon leader. We all expected to go to Korea. I had buddies in Korea, some didn't make it home. The West Point class of 1947 was absolutely decimated. They lost so many 2nd Lieutenants in Korea the class of '47 doesn't even hold reunions. Not enough of them left alive. Out of the 97 junior officers I trained with, 95 went to Korea and the other two were sent to Germany. I was one of the two."

Kronach, Germany: Grimes joins the combination police/military unit known as the Constabulary. Grimes explains, "After WWII there were thousands of 18 and 19-year-old American soldiers with nothing to do. That caused problems. General Patton once stated, 'We need to show the Germans who won the war.' He didn't mean as retribution but to get the troops back to a 'spit and shine' mentality. The army got the boys home and the Constabulary worked with the German police to enforce the law and protect a 1200-mile long border."

Grimes recalled the expectations during the Korean War and Cold War. "We knew the Russians had advisors in Korea and we expected them to come at us in Germany. In truth, we were expendable. My wife kept a packed suitcase in the car at all times. In case of 'emergency', she was to head for the Rhine River; that's where the Allies had chosen to hold the line."

The Constabulary wore helmets with a big yellow 'C' on front ringed by a yellow circle. They were known as the Circle C Cowboys. The Circle C Cowboys could hit 80mph on the autobahn in their M-8 armored cars, with a 37mm cannon and .50 caliber machine gun. Grimes said, "We drove jeeps, too, with the windshield down. We installed a piece of steel on the hoods to snap the piano wires that the East Germans and Russians strung across roads. By the way, Germany has two seasons, the 4th of July and winter. Very cold. Anyway, I recall one night we had about 100 rolling stock railcars on our side of the border. The Russians and East Germans laid track, snuck across the DMZ with a locomotive and stole every one of the railcars." Having served a required three years in a combat branch, Grimes headed back to the states as a 1st Lieutenant. His silver bars were pinned on his shoulders by the legendary airborne officer, General James Gavin.

Fort McClellan, AL.—Grimes recalled, "With a degree in chemistry I guess I was destined for the Chemical Corps, that's the NBC, nuclear, biological, and chemical outfit of the US Army. Guess they liked what they saw because I became an instructor until 1957." That same year the Army wanted Grimes to earn a Master's Degree, at Georgia Tech. He recalled, "I told them they couldn't do that to a UGA graduate; it just wasn't kosher and bordered on cruel and unusual punishment." Told that's where his paycheck would be sent, then-Captain Richard Grimes stomached two years of Georgia Tech and in 1959 received his Master's Degree in Management.

Back to Fort McClellan, AL.—Grimes explained, "We tested chemical warfare systems, safety, always safety, first. If you can manufacture beer, you can manufacture biological agents; if you can manufacture paint, you can manufacture nerve gas. Nerve gas was discovered in Germany. It's a good thing the professor left good notes because the next morning everybody in the lab was dead."

Grimes worked with numerous NBC agents, including Tularemia (rabbit fever), mustard gas, nerve gas, and anthrax. On anthrax: "My father-in-law worked for the US Government at a meat packing plant. If a worker contracted anthrax, they would just take him home, wish him luck, then plan for his funeral. Anthrax is a bovine disease from cattle. People are not immune to anthrax. It's deadly; a couple spores can kill a person."

Summer, 1963: Then Major Richard Grimes is ordered to Vietnam as a chemical advisor to the 2nd Vietnamese Division and for his rendezvous with a defoliant called Agent Orange. He recalled, "I was quartered in Da Nang but worked mainly with remote outposts. We fortified them with barrels of napalm. Soldiers will come through machine gun fire, claymore mines, hand grenades, but not flame. It's a great deterrent."

The program called Ranch Hand: "Ranch Hand was Agent

Orange, plain and simple. The delivery systems included artillery shells, generators to push the chemical downwind, but usually delivered from spray tanks on choppers or aircraft. I remember we were aboard a C-123 less than 100 feet from the ground, the back ramp down, the nozzles spraying Agent Orange, and my buddy lays down on the ramp and tells me to hold his feet so he can take pictures of the chemical being sprayed. Well, I did, but an enlisted guy said, 'Major, that's not going to do you any good.' I think he was telling me if my buddy slid out, so would I."

Spraying Agent Orange in Nam

Grimes cautiously defended the use of Agent Orange. "The side effects, of course, turned out to be horrible, but Agent Orange also saved a lot of lives. We could defoliate the areas around old French fortifications littered with land mines covered by foliage. Many outposts bordered dense jungle; the defoliation saved them from being overrun. It denied the VC crop production and concealment."

Grimes returned home in May of 1964. A full-accounting of Richard Grimes' service must include: the Defense General Supply Center in Richmond, VA, Command and General Staff College at Leavenworth to include a promotion to Lt. Colonel, a stretch at the Pentagon in his own words as an "action officer, not a dang horse holder" (General's Aide), assignment as an exchange officer in England to serve in Her Majesty's Service to teach and lecture at Winterborure, Sandhurst Royal Military College, and prestigious Cambridge University.

He recalled, "Many places had signs posted 'For U.K. Eyes Only' which caused a lot of British officers to ask, 'Why is this Yank teaching us?' I enjoyed England, good people, good soldiers."

Lt. Colonel Richard Grimes retired in 1971 after serving as the Director for Services at the Army Depot at Fort Gillem, GA. "We also oversaw the repair of choppers returned from Vietnam," he said, "The horror of war was evident by dried blood, even body parts. But, we did manage one emergency…the Army was actually running out of the canvas sand bags in Vietnam. So, we are credited with solving the nation's first, and hopefully last, national sandbag emergency!"

Humor has served this patriot well. He suffers from neuropathy in his legs and arms and easily loses his balance. His appendages go numb so badly it's difficult for him to thumb through a book or newspapers or even count paper money. Grimes is on 100% disability from the effects of Agent Orange.

Lt. Col. Richard Grimes

He taught economics and accounting at Clayton State University before completing a 21-year career instructing at DeKalb College. He still does taxes for a few 'friends'.

His closing comments: "Like nuclear warfare, a chemical or biological exchange will hopefully never be needed. I'm a chemist, trained and exposed to what these weapons can do. I came home one day from a field exercise when I was at Fort McClellan and took off my clothes in the laundry room. Well, I must have been careless when on the field. My wife stepped on the pile of clothes and got a big blister on her heel from mustard gas. Not a pretty sight, and not a happy wife, I may add."

The couples' son, Greg Grimes, retired as a 'full-bird' Colonel, one rank above Lt. Colonel. Asked if his son ever needles his father about being outranked, Lt. Colonel Richard Grimes produced a broad grin and said, "Yeah, every chance he gets."

SMOKE FOR BREAKFAST

The annual Atlanta Warbirds Weekend supported by the Dixie Wing's Commemorative Air Force exceeded expectations in 2017 at PDK (Dekalb-Peachtree Airport). Tens of thousands attended the event in celebration of the 75th anniversary of the AVG (American Volunteer Group), better known as the famous Flying Tigers. Their legendary shark-faced nose P-40 Warhawks outgunned and outflew the nimble Japanese Zeroes in occupied China before America entered WWII. The Warbird Weekend agenda was the largest gathering of airworthy P-40s since 1954, and I had the honor of meeting two members of the original Flying Tigers who were in attendance.

Another legend in attendance was Lt. Col. Richard 'Dick' Cole. Speaking with this gentleman was talking to history. On April 18, 1942, 16 heavy B-25 medium bombers took off from the aircraft carrier USS *Hornet* to bomb Japan, thus giving America a much needed boost in morale. The Army pilot who planned and led the mission, Col. Jimmy Doolittle, was piloting the first B-25 to launch. His copilot was Richard 'Dick' Cole. Of the 80 men known as The Doolittle Raiders, Lt. Col. Richard 'Dick' Cole is the only living survivor at 101 years of age.

Incredible stories at Warbird Weekend strolled the tarmac, touched the metal of planes they once flew into combat, and patiently signed books as long lines of admirers and historians bid for attention. And among a plethora of vendors and visitors and veterans, sat Captain Brian Settles in the shade of a tent hawking his book, "*Smoke for Breakfast, a Vietnam Combat Pilot's Story.*" Born a mixed-race child, Settles was a mere 7 days old when given away to an orphanage. As a young man, he piloted the F-4 Phantom jet fighter in an unpopular war. This is his story.

Just seven days old, only one week in this old world before the little boy was given away to the Lincoln Nebraska State Orphanage. He would remain in the orphanage for almost three years until a couple from Muncie, Indiana heard about a mixed-race child up for adoption in Nebraska.

Settles related his early childhood: "My parents got on a train and came out to Nebraska to adopt me then took me back to Muncie. I struggled with feelings of inadequacy, of not being good enough, because of the abandonment, then the adoption. My mother was wonderful. She was a librarian which is a good thing since I developed a love for books and reading. Children today don't have a love for reading and that's a sad commentary on our society."

Captain Settles

His mother was the catalyst, talking her son into attending Ball State University after an athletic scholarship to Colorado fell through. The choice of Ball State channeled his future in aviation. Settles joined the drill team by enrolling in the Air Force ROTC program, attended the ROTC flight instructor's agenda, and after graduation in August '66 entered the pilot undergraduate training program.

Settles still remembers his dreams of pro sports: "I figured the winning formula for success and acceptance would be basketball and football but back to back knee injuries shattered my hoop and gridiron dreams. I keep telling people that I traded my basketball and football uniforms for a fighter pilot's G-suit."

On his decision to fly fighters: "I'm repeating myself, but I kept feeling that I simply wasn't good enough and that I had to prove that I was. So, I chose to fly the Phantom, feeling I really had the fighter pilot's spirit within me, but part of that spirit also reminded me I had to prove my worth, that I was good enough, bad enough, and courageous enough for anything the macho guys could do since fighter pilots are the pilot's pilot sort of thing."

Pausing a moment, Settles clarified, "Let me explain that, fighter jocks don't look down on other pilots, it's just part of the image. I flew 199 missions in Vietnam. I loved flying the F-4 Phantom. I was one of the last back-seat pilots before they transitioned to all back-seat navigators at the end of my tour. Since I was right out of pilot's training, a copilot's job was great. I gained aviation experience, flew wingtip to wingtip formations, and experienced air to air refueling. Landings are a little tough from the backseat, but I loved it, but I didn't love the war enough to volunteer to go back to have a front seat assignment."

On his Vietnam service: "I was stationed at Da Nang in I Corps with the 366th Tac Fighter Wing from August of 1968 until the completion of 199 combat missions. We were known as

'The Gunfighters', but President Johnson had shut down North Vietnam to bombing, which blocked our chances for air to air combat. We did escort recon planes over the North but could only fire when fired upon. Other than that, most missions were air to ground bombing and strafing in South Vietnam."

Asked if he recalled a special mission, Settles replied, "Flying the F-4 was a special mission. We sat on alert duty with a Phantom cocked and ready to go at a moment's notice. When 7th Air Force called to notify us that Marines, Army, or ARVN (Vietnamese Army) were in peril of being overrun, we'd be airborne immediately. Those were special missions for us."

Troops were appreciative. "When I was a Special Duty Officer, Marines would wander into the squadron and ask to meet with Phantom pilots. Those young Marines said they wouldn't be alive today had it not been for the F-4s that made a particular strike. That was a very meaningful experience for us, knowing that what we did as Phantom pilots helped those boys come back home."

His scariest missions? "Search and rescue missions, always. When we had downed pilots and crewmembers that needed to be extracted and rescued as soon as possible, well, things could get a little hairy. Those search and rescue exercises were most likely the only time that the general rules of operating your aircraft could be suspended. Just about anything was in order to rescue those guys. We could get down in the weeds to assist downed airmen, to suppress the 'bad guy' activity; we were allowed to do what needed to be done. Usually we had to follow the rules of engagement, but with a search and rescue, well, it was 'Katie bar the door' stuff."

Settles with his F-4 Phantom

Settles completed 199 combat missions. "After Nam, I flew the KC-135 tankers. I was offered an Air Force Academy appointment but turned it down for employment with Eastern Airlines as a commercial airline pilot. I flew with Eastern for 15 years until they were engulfed in strikes and shut downs and bankruptcy. On a lark, I took a job driving a taxi for what I thought would be about six months. Well, that turned out to be a thirty-month nightmare, driving 300 miles a day in Atlanta traffic, six days a week for less than a third of what I made as a commercial airline pilot. I was a single parent trying to keep two teenage sons off the streets, you do what you have to do."

Settles was finally reemployed in commercial aviation. He retired at 60 years of age as required by the then old-fashioned rules on pilot retirement. Among several successes, including earning a Master's Degree in International Relations from the University of Southern California, Settles served as the Chair and Assistant Professor of Airway Science at Delaware State, is still an active member of Kappa Alpha Psi, the Airline Pilots Association, Retired Airline Pilots Association, and Organization of Black Airline Pilots. He enjoys writing, fishing, his church affiliation, and seeking God's purpose in his life, and teasing his grandchildren.

Captain Brian Settles grew up as a mixed-race child in an era of inequity and exclusion, not to mention personal isolation along with feelings of inadequacy. He defeated the social and personal obstacles, became a Top Gun, earned a Master's Degree, and in doing so has given new meaning to the shabby categorization of 'mixed race', that is to recognize 'mixed race' as the best of both races and cultures. We are all God's children.

THE GULF WAR

"America has fought five wars since 1945 and has gained its objectives in only one of them, the Gulf War."

—Henry Kissinger

LAND MINES, MEMORIES, AND MARINES

Doug Hinton's kinfolk are as southern as cornbread soaking in a glass of buttermilk. The Hintons settled rural Rockdale County, GA in the 1800s and Doug's parents and grandparents rest in peace at Green Meadows on Hwy 138. He lost a great-uncle on Iwo Jima and his great-grandparents are interred at the long standing but pretty much long forgotten Eastview Cemetery on the fringes of postage-stamp-sized Conyers. Civil War relatives rest in peace at the old Smyrna Presbyterian Camp Ground, but Doug's new bride, Cindy, well, the young lady was born and raised in Yankeetown, FL. Go figure.

A 1985 graduate of Heritage High School, Hinton received an appointment to the Merchant Marine Academy from Georgia Senator Sam Nunn. He said, "ROTC at Heritage prepared me for the Academy but within two years I decided on another path."

That 'path' was attending Valdosta State with his first wife. "I needed a degree to receive a commission in the Marines but soon came to realize a military life would not be conducive to a family life. Therefore, I chose the Marine Reserves in 1988."

August 2, 1990: Saddam Hussein's Iraqi Army invades their tiny neighbor, Kuwait. The international reaction entered history as The Gulf War. One week before Thanksgiving, Hinton's unit, C Company of the 8th Marine Tank Battalion, received orders to report for active duty. He said, "Our family Thanksgiving wasn't a very happy affair since I had to report for duty the following Monday."

Sent to Camp Lejeune, NC for deployment, Hinton stated, "It was the largest single gathering of Marines since WWII, 23,000 Jarheads getting primed and ready for a fight." His next port-of-call, via commercial airliner: the Saudi Port City of Jubail (Al Jubayl) as part of the 2nd Marine Division.

No alcohol, no girly magazines, leave the women alone, and be discreet with your Bibles. Doug Hinton and the Marines were definitely on foreign soil. He recalled, "The first thing we did was repaint our equipment desert camouflage since everything was still painted green and stuck out like sore thumbs."

Hinton was assigned a huge wrecker with a big boom on back, used for retrieving immovable or battle-damaged vehicles and to offload tank ammunition. Asked if he'd received specialized training to unload munitions, he replied, "Nope." Asked if his unit had training in desert warfare, he replied, "Nope."

Christmas Eve, 1990: The Marines begin to move north toward the Kuwaiti border. The sand is rock hard, the temperature a surprisingly balmy 80 degrees. The nights turn nippy, rain and hail pound the barren desert; Marines don jackets in January to break the chill.

Hinton durng Desert Storm

January 17, 1991: The air war commences. Hinton stated, "We saw the aircraft overhead as they approached the border with their running lights on. When they reached the border, those running lights disappeared and you only heard the thunder of jet engines. The term 'rolling thunder' comes to mind."

One week later: "It was a bright, clear day. We started hearing rumbles, louder and louder. Over the horizon, we saw a group of B-52s approaching. Those guys were really low. They crossed the border, did a U-turn, then released their bombs. We were almost 10 miles away and felt the ground shake for over a minute."

February 23: The ground war began. The Leathernecks in Hinton's outfit surged into battle with the M-60 tank as their principal armored weapon. The M-60s were Vietnam era weaponry and totally outclassed by the newer and deadlier M-1 Abrams tanks. In military lexicon, the old M-60s were expendable. An expected 70% casualty rate from mines and artillery was to be expected and accepted.

4:45am: Hinton and a two-man crew sat in their wrecker awaiting orders to move forward. They were given three boxes of MREs (Meals Ready to Eat) with 24 or 36 meals per box since the fog of war can isolate troops without resupply. Hinton recalled, "We moved out through the berms into the first mine field. We were so nervous we ate a whole box of MREs."

Armed Humvees, tanks, and troop trucks moved into the jaws of war. A tank hit a mine. Hinton said, "It blew the track off but thank the good Lord the crew was okay. The first row of enemy mines started about three miles beyond the border. You could see them sticking up."

After moving three more miles, the convoy hits the second minefield: "The barbed wire and mines were really thick. You could see mines planted about five feet apart. We had to wear MOPP suits (chemical suits), just in case. We couldn't take any chances.

One of our trucks supposedly detected gas but that was never confirmed. Thank God they didn't use the chemical weapons against us."

Iraqi artillery suddenly opened up on their convoy. Hinton said, "A chopper came up, let loose a salvo of rockets and machine gun fire, and, well, 'All's quiet on the Western Front' so to speak." Iraqi artillery again fired on the Marine convoy, then something incredible took place: The Iraqis began surrendering in droves. "We couldn't believe it," Doug said. "We'd get a token round fired in our direction, then they'd give up, hundreds of them."

Hinton loading ammunition

The vaunted Iraqi Army was half-starved and pitifully ill-equipped to face the Allied Coalition. Hinton recalled, "Some of their soldiers were still wearing street shoes. Those poor fellows had been hijacked off city streets to fill the ranks. Our tanks could no longer stop for POWs, I mean, there were just too many, so we were the ones dealing with prisoners. Strange thing was, the Iraqi soldiers didn't act frightened but grateful, grateful to have been captured and grateful they were getting good food. They'd fight over a pack of Kool-Aid."

Writiing on tank in Desert Storm says it all

After one day of combat, Hinton and his buddies were guarding over 400 Iraqi prisoners. "We had M-16s but we captured tons of AK-47s so we had them as back up, just in case. Later that night, the wrecker became top priority. The Army had swung to the west for a flanking maneuver and ran straight into a minefield. One of their Humvees went up in smoke and fire so we were dispatched to retrieve the vehicle. The Humvee driver didn't make it."

As they moved steadily towards Kuwait City, Hinton and his unit witnessed the horrors of war. "We saw the infamous 'Highway of Death' and other things that made us glad that we were not on the receiving end of our awesome firepower. Thousands of Iraqi soldiers died, but that is war, it's ugly, and it's best to end wars quickly."

Saddam Hussein's 'Mother of all Battles' lasted 100 hours; his Army totally defeated, his military hardware reduced to piles of junk. Hinton remained in Kuwait City for several weeks to assist the inhabitants' return to normalcy, then eventually flew home from the same port city.

Hinton remained in the Reserves for eight years. He earned a degree in Civil Engineering Technology and now works for Corporation Environmental Risk Management as a registered professional engineer.

Doug and Cindy spent their honeymoon as Guardians of WWII veterans on an Honor Flight to Washington, DC. When asked why, Hinton replied, "My war was over in a few months. Some of the veterans on the Honor Flight stayed overseas in WWII for years. It's the least Cindy and I could do for these guys."

GLOBAL WAR ON TERROR

"The purpose of terrorism lies not just in the violent act itself. It is in producing terror. It sets out to inflame, to divide, to produce consequences which they use to justify further terror."
—Tony Blair

DEER CAUGHT IN THE HEADLIGHTS

The morning was typical for the intermodal business, three drivers late for work and three more late for delivery appointments. The railyards were trying to find composure in the chaos as cantankerous owner-operators waited impatiently for lift cranes while complaining about the cost of diesel fuel. The Birmingham line lit up. Birmingham, four drivers virtually on their own, never called early in the morning unless something was wrong.

I took another sip of cold coffee then answered the phone. My senior company driver in Birmingham, Ron, was on the line to report an accident. I grabbed a blank accident report then said to Ron, "Okay, buddy, what happened?"

He replied, "You know that section of the rail yard where you cross the tracks and it's almost impossible to see a train coming around the tight curve? Well, I didn't see it."

Trucks always lose a confrontation with a train, but my driver's welfare came first. "Are you okay?"

"Yeah, I'm fine," he said. "The tractor is okay, too."

Meaning, the rail container was not. "How badly damaged is the rail container?"

"Well, it's more destroyed than damaged."

I suddenly wished my cold coffee was spiked heavily with Kahlua. "Destroyed?"

"Yeah, the container was sort of cut into two big pieces and the cargo is spread all over the railyard, all 45,000 pounds of it."

"What type of cargo?"

"Ragu Spaghetti sauce."

The mental picture was a vision from hell. One of my dispatchers interrupted, "Pete, your wife is on line two. She said it's important."

I pushed the button for line two. "Hey, what's up?" Joyce informed me an airplane had just crashed into one of the twin towers. The morning was going to get a lot worse for a lot of people.

Three hours before American Airlines Flight 11 slammed into the north tower of the World Trade Center, a device called a 'Random Event Generator' at Princeton University projected that a catastrophic incident was about to happen.

Three hours after a machine predicted an upcoming disaster, American Airlines flight attendant Madeline Sweeney was on the phone relaying the unfolding events with a flight services manager when she uttered the last words from Flight 11, "Oh my God, we are way too low."

Thus began a 9/11 tragedy that robbed 3,051 children of a parent and 17 unborn babies of a father. Islamic terrorists had struck America in well-coordinated attacks, using fully-loaded California bound passenger jets heavily laden with jet fuel. In the debris of the towers lay the remains of a Twin Tower Muslim prayer room and a storage area filled with copies of the Koran. Following the attacks, the Ahmadiyya Muslim Community donated 11,170 pints of blood to help the victims of 9/11. Ascertaining the good guys from the bad guys is still a challenging decision, for both sides.

After Joyce notified me of a second plane hitting the second tower, I knew we were under attack. I left my Special Service Division Office and entered the domain of the over-the-road and inter-city intermodal divisions. Standing by the door, I informed the 20 or so employees busily dealing with freight issues, "People, a second plane has hit the World Trade Center. Our country is under attack." Deer caught in the headlights. Stunned, skeptical, no emotion, and no reply. I shut the door and returned to my office.

A few years later, I interviewed four eye-witnesses to 9/11 for my newspaper article "A Veteran's Story." They experienced a wide range of roller coaster emotions from fear to frustration to revengeful fury.

Army veteran Terri Prieto was working in the east side of the Pentagon when American Airlines Flight 77 hit the west side of the building. She didn't even feel the impact. Four of Terri's friends were among the 125 civilian and military personnel killed. Her supervisor received a call and was told a helicopter had hit the west E ring. He bolted out the door to safety, then the security officers began emergency evacuations. Outside in a grassy area opposite the parking lot, Terri soon realized the aircraft that smacked into the Pentagon was much bigger than a helicopter. "I saw large pillars of thick smoke billowing from the west wing," she said. "Then fighter jets arrived overhead and security officers kept shouting, 'Move back! Move back! Another plane is inbound!' That other plane was the one that went down in Shanksville, but I doubt if anybody that day knew where it was headed."

Told to go home, Terri fought traffic jams, picked up stranded survivors, and tried to call family members on her cellphone, but all the airwaves were overloaded. She finally got through later that evening.

On 9/12, she was back at work. "I was proud to see the American flag draped over the side of the Pentagon," she said. "It was our way to say to the bad guys, 'you only won the first round.' We noticed the vending machines had been vandalized but later found out the first responders had broken into them for soft drinks and to keep the survivors hydrated. Oh, by the way... one of my friends saw the airplane hit the Pentagon, so those conspiracy freaks need to get a life!"

Stress and nightmares still plague Terri. "It's getting easier," she stated. "But I was there, I saw it, smelled it, breathed it, and I'll never forget it. I hope my country doesn't either."

Price Waterhouse Cooper employee Mike Casillas stated, "I had arrived for work when the first plane hit so I joined about 50 other people on the waterfront staring at the North Tower. Smoke was pouring out of a big hole in the side of the building. I called my wife and while we were talking the second plane came around and I saw it crash into the South Tower. I knew immediately we were under a terrorist attack. We were all hoping and praying for the people trapped in the buildings, debris was falling from windows, then I saw a person jump."

We all became veterans on 9/11

Mike was emotionally shaken after seeing the jumper. "That really upset me," he said. "Then the first building came down. I was heartsick and felt like throwing up, knowing all those people in the building were dying."

Mike, like most people, wanted to help. "After the second tower fell, a huge dust cloud blocked our vision so a buddy and I found a hotel filling with survivors covered in dust. The hotel manager sent employees out to buy food for the survivors, even cat and dog food for their pets. I'll never forget 9/11. Had I arrived for work early that morning, you and I wouldn't be talking. My shock turned to anger. You see, one of our co-workers and his two-year-old kid were on one of those planes. As far as I was concerned, I wanted our military to go over there, find the guilty people, then nuke the place into a plate of glass."

Meanwhile, Flight 77 hit the Pentagon at 9:37am. The only American not on Mother Earth was astronaut Frank Culberston circling above the tragedy aboard the International Space Station. Notified of the disaster unfolding, Culberston started recording the event from outer space. He found out later the commercial pilot of American Airlines Flight 77 was Charles Burlingame, his classmate from the Naval Academy.

At 9:45am, the US Capitol and White House employees started evacuating their buildings, eyes glazed like deer caught in the headlights. Less than 20 minutes flying time to their fourth target, the terrorists in command of United Airlines Flight 175 were fighting to die while the travelers aboard were fighting to live. Having said 'goodbye' to his wife, passenger Todd Beamer left his cellphone on and could be overheard urging his fellow passengers, "Let's Roll." Minutes later, the plane disintegrated into a million pieces as it slammed into the ground at 500 mph in Shanksville, PA at 1003am. Only temporarily blinded by the headlights, the passengers aboard Flight 175 saw the light then fought back like tigers with their eyes wide open. After the crash, Todd Beamer's cellphone remained on until it too died fifteen minutes later.

Bill LeCount recalls evacuating the PWC building less than

a half mile across the Hudson River from the Twin Towers. "We had made it out to the evacuation area when the second plane hit the South Tower. The explosion was enormous; a gigantic ball of flame. We were all in a state of shock. Later I took photos of the boats going back and forth to bring people to safety. It was an incredible act of human kindness. The boats transported over 500,000 people from Manhattan. One of the saddest things was watching hundreds of refrigerated trucks passing by to load the projected number of bodies to be brought across."

Second plane approaching South Tower

Bill's co-worker, Martin Burton recalled, "It took a while for everything to sink in. It was all so unbelievable. We went to Bill's apartment just as the South Tower came down. It caused a thunderous rumble, easy to feel and hear from half a mile away. I felt like we were in a bad movie, then the second tower came down. I went numb, like detached from the whole thing. For me, the real horror set in about two or three days later, of how cold and calculating that attack was on 9/11."

The towers were gone. The Pentagon burning. A big Boeing 767 reduced to confetti in an isolated field in Pennsylvania. I went back to the intermodal and over-the-road room to check on folks. Except for conducting what was now limited business on the phones, everyone remained silent as they tried to cope with the impossible becoming reality. They reminded me of deer caught in the headlights.

One hundred days after the 9/11 attack, New York firefighters finally extinguished the last fire in the rubble of what used to be the Twin Towers. Of the thousands who perished, only 291 bodies were found 'intact' at Ground Zero, yet recovery efforts discovered 437 watches and 144 wedding rings. The recovery efforts through DNA continue to this day. A male was officially identified this year.

The terrorist group, Al Qaeda, declared war on the United States via Osama Bin Laden in 1996. After a string of successful terrorist attacks against American targets versus unsuccessful if not feeble retaliatory strikes from the U.S., the desert rat roared on 9/11/01. Expressions on the faces of George W. Bush down to the most junior Congressman said it all: deer caught in the headlights.

As of this writing, North Korean intimidator Kim Jong-un

is rattling his tiny nuclear sword in the face of America and her South Pacific allies. Kim Jong-un has ordered his troops to 'be ready at any time to strike' and American intelligence sources believe it's likely he'll launch a missile to 'land in the vicinity' of Guam. The North Korean news agency has urged all nations, 'not to make hasty or war decisions based on an unintended hit.' Think about that statement. Have you noticed the strange expressions on the faces of world leaders lately?

We are all veterans and victims of 9/11. Let's pray we're not caught in the headlights a second time, because this struggle will not be over for a long, long time. Nor will the stories.

About the Author

Pete Mecca is an Air Force intelligence veteran of Vietnam, having served two and a half years in Vietnam. His main love is working with veterans from all wars. His award-winning weekly full-page article "A Veteran's Story" is featured in several Georgia newspapers. Pete has been featured in Georgia Magazine, appeared on Atlanta's Channel 69, and worked as a consultant for PBA-30 during the premier of Ken Burn's documentary on Vietnam. Pete serves as commander and program director for the Atlanta World War II Round Table, is an active member of Atlanta Vietnam Veterans Business Association, Golden Warriors, American Legion, the Georgia Vietnam Veterans Alliance, and VFW. He coordinates three monthly veteran brunches and conducts lectures, symposiums, and panel discussions throughout Georgia, including schools. The second book in this Veterans: Stories from America's Best series will be released by Deeds Publishing in the fall of 2018. Contact Pete via his website: www.veteranssfab.com

CPSIA information can be obtained
at www.ICGtesting.com
Printed in the USA
LVOW13s0008090418
572732LV00001B/1/P